Requirements Engineering for Computer Integrated Environments in Construction

Ghassan Aouad
Pro-Vice Chancellor for Research & Innovation
University of Salford, UK

&

Yusuf Arayici
Lecturer and MSC Programme Co-ordinator
Faculty of Business, Law and the Built Environment
University of Salford, UK

WILEY-BLACKWELL
A John Wiley & Sons, Ltd., Publication

This edition first published 2010

Blackwell Publishing was acquired by John Wiley & Sons in February 2007. Blackwell's publishing programme has been merged with Wiley's global Scientific, Technical, and Medical business to form Wiley-Blackwell.

Registered office
John Wiley & Sons Ltd, The Atrium, Southern Gate, Chichester, West Sussex, PO19 8SQ, United Kingdom

Editorial offices
9600 Garsington Road, Oxford, OX4 2DQ, United Kingdom
2121 State Avenue, Ames, Iowa 50014-8300, USA

For details of our global editorial offices, for customer services and for information about how to apply for permission to reuse the copyright material in this book please see our website at www.wiley.com/wiley-blackwell.

Library of Congress Cataloging-in-Publication Data

Aouad, Ghassan.
Requirements engineering for computer integrated environments in construction / Ghassan Aouad and Yusuf Arayici.
p. cm.
Includes bibliographical references and index.
ISBN 978-1-4051-8945-3 (hardback : alk. paper)
1. Communication in the building trades. 2. Building – Superintendence – Data processing. 3. Requirements engineering. I. Arayici, Yusuf. II. Title.
TH215.A66 2010
690.0285 – dc22

2009028107

A catalogue record for this book is available from the British Library.

Typeset in 11/13.5 Sabon by Laserwords Private Limited, Chennai, India
Printed and bound in Malaysia by Vivar Printing Sdn Bhd

1 2010

Contents

Foreword

Collaborative working using innovative integrated Information and Communication Technology (ICT) systems in construction has become a reality. Many activities are now performed in a distributed manner with stakeholders situated in discrete geographical locations. Computer integrated environments (CIE) and building information modelling (BIM) are the sorts of innovative system that help reduce fragmentation and enable construction stakeholders to collaborate on projects.

CIE and BIM have been the subject of research for many years but their uptake has been limited due to a lack of development of the technology and its ineffective implementation. Both industrialists and researchers argue that networking, collaboration, information sharing and communication – all important issues for the future of construction – can be managed through CIE systems. For the technology to develop successfully, appropriate delivery and effective implementation of user and industry-oriented CIE and BIM systems is crucial.

Requirements engineering is a branch of systems engineering and is related to the development of CIE technology and its effective implementation; it helps determine what to develop, how to develop it, and when it should be implemented. Requirements engineering addresses the goals, desired properties and constraints of complex systems such as the CIE systems that involve software systems, organisations, people and process. It also covers all activities related to the acquisition, specification and maintenance of requirements throughout the system development lifecycle, as well as how requirements relate to business processes, work redesign, system and software architecture, testing and validation.

This book is about the development of a requirements engineering (RE) framework for CIE system development and BIM implementation. Through case study projects, it presents mechanisms to help run construction projects smoothly and collaboratively from the early briefing stage to the detailed design and on to the end of the construction phase and over the project lifecycle.

The RE framework proposed here is targeted at CIE systems. The key features of the framework are being (1) ready-to-use; (2) simple; (3) domain specific; (4) adaptable; (5) systematic; and (6) integrated with the legacy systems. The method has three key constructs:

- Techniques for requirements development, which includes the requirement elicitation, requirements analysis/modelling
- Requirements validation
- Requirements documentation and facilitating the requirements management

In short, the book focuses on the user-driven CIE system development methodologies and successful BIM implementation and adoption in the construction industry. I would like to conclude this foreword by suggesting that this book will give added value for user-led CIE/BIM development and implementation.

Arto Kiviniemi
Vice President for Innovation and Development
Olof Granlund, Finland

Preface

Computer integrated environment (CIE) is the type of innovative integrated information system that helps to reduce fragmentation and enables the stakeholders to collaborate in business. Researchers have observed that the concept of CIE has been the subject of research for many years but the uptake of this technology has been very limited because of the development of the technology and its effective implementation. Although CIE is very much valued by both industrialists and academicians, the answers to the question of how to develop and how to implement it are still not clear.

Industrialists and researchers conveyed that networking, collaboration, information sharing and communication will become popular and critical issues in the future, which can be managed through CIE systems. To enable successful development of the technology, successful delivery, and effective implementation of user- and industry-oriented CIE systems, requirements engineering seems a key parameter. Therefore, through experiences and lessons learnt in various case studies of CIE systems developments, this book explains the development of an RE framework specific to the CIE system.

The RE framework that has been developed in the research is targeted at CIEs with a particular interest in the construction industry as the implementation field. The key features of the RE framework are that they are (1) ready to use, (2) simple, (3) domain specific, (4) adaptable, (5) systematic and (6) integrated with the legacy systems. The method has three key constructs: (i) techniques for requirements development, which includes the requirement elicitation, requirements analysis/modelling and requirements validation; (ii) requirements documentation and (iii) facilitating the requirements management. It focuses on system development methodologies for the human-driven ICT solutions that provide communication, collaboration, information sharing and exchange through CIEs for professionals situated in discrete locations but working in a multidisciplinary and interdisciplinary environment. The overview for each chapter of the book is given below.

Chapter 1 provides an overview by setting the scene and presents the issues involved in RE and CIE. Furthermore, it makes an introduction to the necessity for RE for CIE system development, experiences and lessons learnt cumulatively from CIE systems developments that the authors have been involved in and the process of the development of an ideal RE framework for CIE systems development.

Chapter 2 aims at building up contextual knowledge to acquire a deeper understanding of the topic area. This includes a detailed definition of the RE discipline and the importance and principles of RE and its process. In addition, state-of-the-art techniques and approaches, including the contextual design approach, the use case modelling and the agile RE processes, are explained to provide contextual knowledge and understanding about RE to the readers.

Chapter 3 attempts to identify a scope and contextual knowledge and understanding about CIE and building information modelling (BIM). In doing so, previous experiences of the authors about systems developments for CIE are explained in detail as the CIE/BIM case studies.

In the light of contextual knowledge gained about RE in Chapter 2, in order to realise the critical necessity of RE to combine technology, process and people issues in the right balance, Chapter 4 will critically evaluate the RE activities of CIE systems developments that are explained in Chapter 3. Furthermore, to support the necessity of RE for human-centred CIE systems development, the findings from semi-structured interviews are shown in a concept map that is also explained in this chapter.

In Chapter 5, RE is investigated from different angles to pick up the key issues from discrete research studies and practice such as traceability through process and product modelling, goal-oriented RE, the essential and incidental complexities in requirements models, the measurability of quality requirements, the fundamentals of RE, identifying and involving the stakeholders, reconciling software requirements and system architectures and barriers to the industrial uptake of RE. In addition, a comprehensive research study measuring the success of RE processes through a set of evaluation criteria is introduced. Finally, the key issues and the criteria are comparatively analysed and evaluated in order to match each other and confirm the validity of the criteria for the evaluation and assessment of the RE implementation in the CIE case study projects in Chapter 7 and the key issues will be used in Chapter 9 to support the Capability Maturity Model (CMM) for acceptance and wider implications of the RE framework to be proposed in Chapter 8.

Chapter 6 explains and particularly focuses on how the RE activities in the case study projects were handled, by highlighting the strengths and weaknesses. This will also include the experiences and lessons learnt from these system development practices. The findings from these developments will also be utilised to support the justification of the necessity of an RE framework for the CIE systems developments. In particular, the following are addressed:

- Common and shared understanding in RE efforts
- Continuous improvement
- Outputs of RE
- Reflections and the critical analysis of the RE approaches in these practices.

The premise of Chapter 7 is to evaluate and assess the RE approaches in the CIE case study developments from multiple viewpoints in order to find out the strengths and the weaknesses in these RE processes. This evaluation will be mainly based on the set of criteria developed by the researchers and developers in the RE community in order to measure the success rate of the RE techniques after their implementation in the various system development projects. This set of criteria has already been introduced in Chapter 5. This critical assessment includes conducting a questionnaire-based survey and descriptive statistical analysis.

In Chapter 8, the RE techniques tested in the CIE case study developments are composed and compiled into an RE process in the light of the strengths and weaknesses identified in the previous chapter through benchmarking with a CMM to ensure that it has the required level of maturity for implementation in the CIE systems developments. As a result, a framework for a generic RE process for CIE systems development will be proposed.

In Chapter 9, the authors will discuss the acceptance and the wider implications of the proposed framework of RE process using the CMM from Chapter 8 and the key issues from Chapter 5.

Chapter 10 is the concluding chapter and it summarises the findings and brings the book to a close with recommendations for the implementation of the proposed RE framework and also prescribes a guideline as a way forward for better implementation of RE for successful developments of the CIE systems in the future.

Acknowledgements

This book has been prepared from the experience and lessons learnt and research outcomes in various CIE developments in Salford such as nD Modelling and DIVERCITY. In every development, soft issues became more and more important. In other words, there has been a transition from testing and demonstrating the concepts such integrated computer environments and building information modelling towards the need of the stakeholders in construction and practicality and scalability of those CIE developments.

As a result, this book is produced from the PhD study by Dr Yusuf Arayici under the supervision of Prof. Ghassan Aouad by looking into RE practices and experiences in a number of the CIE developments such OSCON, GALLICON, DIVERCITY and nD Modelling to specifically identify an RE framework for the integrated, collaborated, distributed VR-based system and BIM implementations so that this study can contribute to the future CIE systems developments and BIM implementations projects to ensure successful deployments. Therefore, we would like to express our special thanks to the OSCON, GALLICON, DIVERCITY and nD Modelling consortia.

As an academic effort, this book has been developed at the School of Built Environment at the University of Salford. We also would like to thank Lynn Williamson who constantly helped us edit the book.

Abbreviations

AEC	Architecture, Engineering, Construction
ATLAS	Architecture, Methodology and Tools for Computer Integrated Large-Scale engineering
BIM	Building Information Modelling
CAD	Computer-Aided Design
CBSP	Component-Bus-System-Property
CIC	Computer Integrated Construction
CIE	Computer Integrated Environments
CIM	Computer Integrated Manufacturing
CMM	Capability Maturity Model
CORBA	Common Object Request Broker Architecture
COTS	Commercial Off the Shelf
DIVERCITY	Distributed Virtual Workspace for Enhancing Communication within the Construction Industry
GenCOM	General Construction Object Model
GQM	Goal/Question/Metric
IAI	International Alliance for Interoperability
ICON	Information/Integration for Construction
IDEF0	Integration Definition for Function Modelling
IFC	Industry Foundation Classes
ICT	Information and Communication Technology
JAD	Joint Application Development
OPIS	Object Model-Based Project Information System
OSCON	Open Systems for Construction
R&D	Research & Development
RD	Requirements Documentation
RE	Requirements Engineering
REAIMS	The Requirements Engineering Adaptation and Improvement for Safety and Dependability
ROI	Return-on-Investment
RUP	Rational Unified Process
SME	Small and Medium-Size Enterprises
SOAP	Simple Object Access Protocol

SPACE	Simultaneous Prototyping for an Integrated Construction Environment
STEP	Standard for the Exchange of Product Model Data
TTM	Time to Market
UCDA	Use Case-Driven Analysis
UED	User Environment Design
UML	Unified Modelling Language
VE	Virtual Enterprises
VR	Virtual Reality
VRML	Virtual Reality Modelling Language
WISPER	Web-Based IFC Shared Project EnviRonment
XML	Exchange Mark-up Language
XP	Extreme Programming

1

Introduction

1.1 Definitions

1.1.1 Computer integrated environments

A computer integrated environment (CIE) is used to establish an alliance of enterprises that come together to share skills or core competencies and resources in order to better respond to business opportunities, and integrate people whose cooperation is supported by computer networks. It is a manifestation of collaborative networks and a particular case of virtual organisation (VO) or virtual enterprise (VE).

Collaborative working using innovative CIE systems has become a reality as many activities in multidisciplinary work environments are performed with the stakeholders situated in discrete geographical locations. Such an information and communication technology (ICT) system helps to reduce fragmentation and enables the scattered professionals to communicate and collaborate virtually together in a synchronous or asynchronous manner.

The concept of CIE has been the subject of research for many years but the uptake of this technology has been very limited because of the development of the technology and its effective implementation. However, professionals and researchers have confirmed that networking, collaboration, information sharing and communication will be significantly more crucial and in demand in the future as conducting business in multidisciplinary and interdisciplinary environments will be essential for success and sustaining competitiveness. Subsequently, for successful development of the CIE technology as well as for the delivery and effective implementation of user-oriented systems, which require a correct balance between technology, process, organisation and people aspects, requirements engineering (RE) has become critically important. The specific features of CIEs or VEs can be summarised as follows:

- Boundary crossing
- Complementary core competencies

- Geographical dispersion
- Complementary nature of the partners
- Participant equality
- Extensive use of information and communications technology.

Over the last couple of decades, there has been a major shift from an industrial economy to that of an information economy. This has led to an enormous increase in competitiveness among companies, and new technology is needed to help capitalise on the information economy. CIEs for VEs or VOs is a new and major trend in the cooperative business. CIE allows businesses to specialise and be flexible within their environments. In the past, this business model has been applied to outsourcing and supply chains, as well as to temporary consortia. Owing to the fact that the formation of these VEs is an intricate process, a new form of technological support has thus been developed. The most ambitious of the support systems actually intends to automate part of the creation process as well as the operation of these enterprises (Cardoso and Oliveira, 2005).

As with all types of enterprises, VEs present both benefits and challenges. Organisations can benefit from VEs through more economical connections with suppliers, greater opportunities to create revenue, more efficient operations and a reduction in administrative costs. The challenges facing VEs are inexperienced users, secured access, expense control and the level of integration and interoperability required to create a successful VE (Sun Microsystems, Inc., 2004).

Adopting VE enables a conventional organisation to accomplish tasks traditionally meant for an organisation that is much bigger, better resourced, and financially stable. One company having the technical capability, another with the right human skill set and a third with the solution, may come together to create a VE. For example, the current situation within the construction industry is that many projects are one of a kind and involve coordination between practitioners such as designers, engineers and suppliers. A typical construction project consists of a number of organisations and teams that are brought together for the duration of that particular project to form a so-called 'virtual enterprise'. This enterprise often contains units that are in different physical locations and use different computer platforms and have a need to work collaboratively and share the same project data (Faraj *et al.*, 2000). Some of the key benefits include, but are not limited to, the following:

- Emphasis on collaborative work for the construction stakeholders. The industry currently suffers from a considerable degree of fragmentation.

- Proposed new data exchange standards, such as Industry Foundation Classes (IFCs), for information exchange among the stakeholders.
- Proposed new construction processes, which eliminate non-value-adding activities.
- Provision of shared access to project information via integration over a central database or a communication layer. This prevents information duplication among stakeholders.
- Claims of providing savings in lifecycle project costs and time.
- Limited use in industry; thus there is little experience on their use.
- Provision of virtual reality (VR) functions and 4D simulations for decision-making processes for optimised solutions.

There have been extensive studies in the last decade such as ATLAS (Greening and Edwards, 1995), COMBINE (Augenbroe, 1995), RATAS (Björk, 1994), ICON (Aouad *et al.*, 1994), COMBI (Scherer, 1995), OSCON (Aouad *et al.*, 1997), OPIS (Froese and Paulson, 1994), SPACE (Alshawi *et al.*, 1997), ToCEE (Amor *et al.*, 1997) WISPER (Faraj *et al.*, 2000), GALLICON (Sun *et al.*, 2000), BIDSAVER (http://www.ceconsulting.it/ve/bidsaver.html), ALIVE (Chris *et al.*, 2001), LEGAL-IST (www.legal-ist.org), ECOLEAD (Lavrac *et al.*, 2005) in the area of CIE.

1.1.2 Requirements engineering

RE is a branch of systems engineering and it is related to the development of the technology and its effective implementation. That is to say, it helps to define what and how to develop, and when to implement the technology. RE is concerned with the goals, desired properties and constraints of complex systems such as the CIE systems that involve software systems, organisations and people.

Furthermore, it covers all activities related to the acquisition, specification and maintenance of requirements throughout the lifecycle of the software development projects. It also covers how requirements relate to business processes, work redesign, system and software architecture and testing and validation. This process is regarded as one of the most important aspects of building an information system as it is during this process that a decision is taken on what is to be built.

RE is also known as systematic requirements analysis (Wiegers, 2003). It is sometimes referred to loosely by names such as *requirements gathering*, *requirements capture*, or *requirements specification*. RE is critical to the success of a development project (Abran *et al.*, 2005). Requirements must be actionable, measurable, testable, related to identified business needs or opportunities and defined to a level of detail sufficient for system design.

Carr (2000) defined properties, attributes, services, function and behaviours as the requirements that are needed in a product to accomplish the goals and purposes of the system to be developed. RE is an iterative process by which the needs and requirements of individuals and groups significant to the product development are researched and identified. RE defines the following (Cooper *et al.*, 1998):

- Customer, user and market requirements
- Design requirements
- Technical requirements.

Maguire (1996) emphasised that adopting a user-centred design process leads to more usable systems and products. It reduces the risk that the resulting system will under-deliver or fail. User-centred design implies the following:

- Early focus on users, tasks and environment
- Active involvement of users
- Appropriate allocation of function between user and system
- Incorporation of user-derived feedback into system design
- Iterative design whereby a prototype is designed, tested and modified.

Numerous surveys have been conducted, which conclude that project failures are caused by a lack of proper attention to requirements processes. A survey, which was undertaken by The Standish and Gartner groups, reports that only 26% of software projects are considered successful while 74% are unsuccessful.

When detailing the causes of success or failure, the most frequent area is the subject of user requirements (Eberlein and Leite, 2002). The CHAOS development report published in 1995 and 2000 by The Standish group showed that almost half of the projects failed or were cancelled because of the lack of RE effort. The main reason for project success for a similar percentage of projects was good RE (Eberlein and Leite, 2002).

Another survey by McPhee and Eberlein (2002) showed that senior software developers and project managers believe that the requirements activities should account for 25% of the total development effort. This was the outcome of the survey although the survey focused on the projects that had a critical time for delivery to the end users. As a result, this survey clearly proves that RE is crucial in systems development. Understanding the users' real requirements is absolutely critical to the development of successful information systems.

To achieve a user-oriented and high-quality system, it is important that the user requirements must be captured and modelled in the right way. If done correctly, the system to be developed will meet the user needs and lead to better user satisfaction and implementation. On the other hand, if the user requirements are not defined correctly, the software is less likely to meet the user requirements, even if the software conforms to the requirements specifications developed.

1.2 Why Requirements Engineering Is Needed for the CIE Development

CIE is an important solution for the integration of the processes through the supply chain. Research has emphasised some of the benefits of integration such as reducing the project lifecycle and cost, removing the non-value-added activities for achieving lean processes and production, encouraging collaboration and increasing client satisfaction (Sun and Aouad, 2000). For example, the construction industry is of a multidisciplinary, traditional and fragmented nature, which results in many different issues and bottlenecks such as lead time, lack of buildability, increased cost, unsatisfied clients and inefficient documentation, to name a few. These challenges can be overcome by implementing CIE systems. However, people in the industry have little awareness of how to use such systems effectively in their work environment because of unfamiliarity with such systems; this results in a gap between the developers of CIEs and industrialists.

One route to developing more user-centred systems is the use of systems development methodologies, which are appropriate to the CIE systems. However, there is currently little debate within the research community as to what the characteristics of a CIE systems development methodology should be.

To date, CIE researchers, especially within the construction industry, have had little focus on RE in systems development, which is actually necessary to develop user-oriented and more practical CIE systems. Despite the increasing interest by both academia and practitioners in CIE, there is little research to identify the best practices in RE. On the other hand, according to the Vision reports published by Aouad *et al.* (1998) and Sarshar *et al.* (2000, 2002), communication, networking, integration and information sharing will be major issues over the next 10 years in the construction industry. The increasing trend towards the implementation of Building Information Modelling (BIM), which enables information sharing, collaboration and interoperability, will make the uptake of the CIE systems inevitable for the construction industry.

BIM use in some countries such as Denmark is a mandated legal requirement in public property projects. It is believed that these legal requirements will soon be in place in other countries including the United Kingdom for the implementation of CIEs-so called BIM-as a result of moving towards a knowledge economy. Consequently, RE will be vital for the successful development of CIE technologies for virtual enterprising such as BIM-based CIEs for the construction industry. Employing appropriate requirements techniques will provide the following benefits:

- More practical CIE systems
- Increased usability and ease of use
- Configurable systems
- Flexible and scalable systems
- Contribution towards closing the gap between the practitioners and the researchers
- Contribution towards increasing the uptake of CIEs by the industrialists
- Support for the business processes modelling and the product modelling.

Requirements engineering techniques and methods can vary according to the nature, structure and size of the system development project. In other words, while an RE technique works well for one type of software system development, the same method may not work well for another type of system. Therefore, it is necessary to define an RE method that is targeted at software systems for CIEs. The method should provide a standard template of the RE process that is applicable to different CIE systems developments. Therefore, RE will be addressed in this book with a particular focus on the following issues.

- Ascertain the level of awareness about RE in the CIE community and justify the need for the identification of a RE framework.
- Gain a deeper understanding of the RE concept in system development.
- Evaluate the RE approaches explored.
- Elaborate the RE approaches in the case study developments.
- Analyse and evaluate experimented RE approaches in the case study projects.
- Master an RE approach for future CIE system development based on the analysis and evaluations.

- Validate and implicate the mastered RE framework for future studies and CIE developments.

1.3 How the Requirements Engineering Approach Is Formulated

Once contextual knowledge about RE and CIEs is established in the book, it is followed with the identification of the assessment criteria for the RE approaches. In order to extract the key issues of RE, discrete research studies that approach RE from different angles are investigated. In order to set up associations between the key issues and the characteristics of an RE process in a CIE development, the relevance and suitability of these key issues for CIE systems development are also considered and discussed.

In order to measure the success rate of the RE processes under evaluation, the book explains an instrument to evaluate or assess the RE processes. This leads to an analysis of the key issues and criteria through a benchmarking process between the two. Such analysis enables confirmation of the validation of both the key issues and the criteria by establishing categories, sub-groups, relationships and possible dependencies (Hammersley and Atkinson, 1983). Figure 1.1 shows the coherent relationships between the groups.

The criteria specified in the book are designed to measure the success of the RE processes after their implementation in system developments. Therefore, using the criteria for the assessment of RE activities in the case studies is more appropriate than the key issues. However, the key issues will be used as part of the validation process of the proposed RE process at the end.

After the criteria are benchmarked with the key issues, the RE carried out in the case study projects is analysed to explore the strengths and the weaknesses according to these criteria. From the analysis of the case studies, strengths and weaknesses and any problems in the RE approaches adopted in the case study, projects are clearly identified.

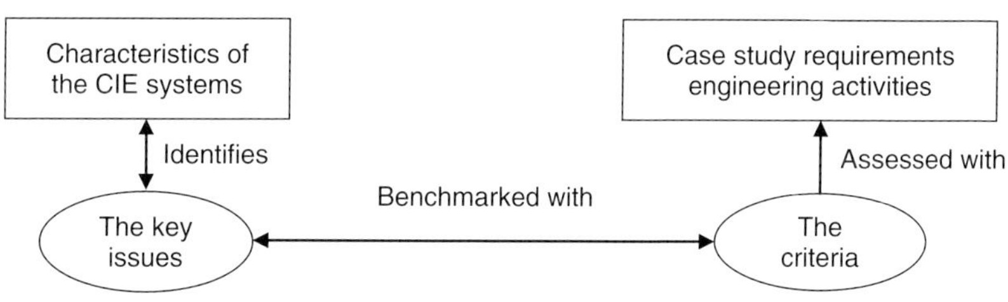

Figure 1.1 Criteria development to assess the requirements engineering experiences.

The process is then enhanced to cure the weaknesses and provide a ready-to-use, simple, adaptable, systemic, domain-specific RE process for the future CIE developments. The enhanced RE framework is validated before proposing it to the CIE community. Because empirical validation is not possible currently, only theoretical validation is conducted comprehensively. For empirical validation, the framework should be implemented in a CIE development project. It is done in two steps – internal (dependent) evaluation and external (independent) evaluation.

Internal evaluation is done against the key issues to be explained later in Chapter 5 of the book. The use of the key issues is more appropriate than the criteria for internal evaluation because the criteria are designed to measure the success of the RE process after the implementation, while the key issues are actually designed to improve the RE process before its implementation. Lastly, making use of the key issues for internal evaluation will allow the setting up of a coherent relationship and a good balance between the stages of the research methodology. This is depicted in Figure 1.2.

Figure 1.2 denotes the evolution of the development of the proposed RE framework. The association between the key issues and the criteria is established through benchmarking analysis. After the critical analysis and elaboration of the RE activities in the case studies, the association between the criteria and the case study RE activities is established in order to evaluate, analyse and measure the success of the RE processes in the case studies. On the basis of this evaluation and analysis, a further association is established between the RE activities in the case studies and the proposed RE framework. A final association is established between the key issues and the proposed RE framework to enable internal validation.

External evaluation is conducted through benchmarks against the external assessment models. Two different models are used for the

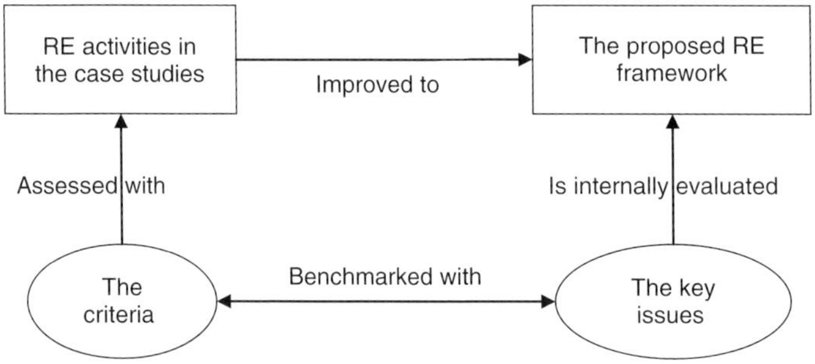

Figure 1.2 Interrelations between different requirements engineering aspects in the development of the ideal RE framework for CIE systems development.

external evaluation the Requirements Engineering Adaptation and Improvement for Safety and Dependability (REAIMS) assessment model (Sommerville and Sawyer, 1997), which is a capability maturity model, and project risk factors determined by Keil *et al.* (1998), Carr (2000) and CHAOS Survey (Standish Group, 1995, 2000).

Lastly, the validated RE process is recommended to be applied to the current or previous CIE development projects and systems to realise what has been underperformed in these developments in regard to RE.

References

Abran, A., Moore, J.W. Executive editors, Bourque, P., Dupuis, R., editors, (2005), "Chapter 2: software requirements", Guide to the Software Engineering Body of Knowledge, 2004 Version, Los Alamitos, California, IEEE Computer Society Press, ISBN 0-7695-2330-7.

Alshawi, M., Putra, C.W., Faraj, I., (1997), "A Structured Concept for Objects Life Cycle in Integrated Environments", Microcomputers in Civil Engineering, 12:339–351.

Amor, R.W., Clift, M., Scherer, R., Katranuschkov, P., Turk, Z., Hannus, M., (1997), "A framework for concurrent engineering–ToCEE", European Conference on Product Data Technology, PDT Days 1997, CICA, 15–16 April, 1997, Sophia Antipolis, France, pp. 15–22.

Aouad, G., Betts, M., Brandon, P., Brown, F., Child, T., Cooper, G., Ford, S., Kirkham, J., Oxman, R., Sarshar, M., Young, B., (1994), "ICON (Integration of Construction Information): Integrated Databases for the Design, Procurement and Management of Construction". Final Report, Department of Surveying and Information Technology Institute, University of Salford, Greater Manchester, United Kingdom.

Aouad, G., Hinks, J., Cooper, R., Sheath, M.D., Kagioglou, Sexton, M., (1998), "An IT Map for a Generic Design and Construction Process Protocol", Journal of Construction Procurement, 4(2):22–44.

Aouad, G., Marir, F., Child, T., Brandon, P., Kawooya, A., (1997), "A Construction Integrated Databases–Linking Design, Planning and Estimating", International Conference on Rehabilation and Development of Civil Engineering Infrastructure Systems, Lebanon. American University of Beirut, Beirut.

Augenbroe, G., (1995), "An Overview of the COMBINE Project, Proceedings". In R.J. Scherer, editor, ECPPM'94: Product and Process Modelling in the Building Industry, Dresden, Germany, Balkema, pp. 547–554.

Björk, B.C., (1994), "The RATAS Project–An Example Of Co-operation between Industry and Research Toward Computer Integrated Construction", Journal of Computing in Civil Engineering, ASCE, 8(4):401–419.

Cardoso, H.L., Oliveira, E., (2005), "Virtual Enterprise Normative Framework within Electronic Institutions". http://paginas.fe.up.pt/~eol/PUBLICATIONS/2005/esaw_post.PDF, Retrieved June 7, 2006.

Carr, J., (2000), "Requirements Engineering and Management: the Key to Designing Quality Complex Systems", The TQM Magazine, 12(6):400–407, ISSN 0954-478X.

CHAOS Report, (1995), "Software Development Report", Boston, Massachusetts, The Standish Group, www.standishgroup.com.

CHAOS Report, (2000), "The Software Development Report", Boston, Massachusetts, The Standish Group, www.standishgroup.com.

Chris, C., Hassan, T., Merz, M., White, E., (2001), "The eLegal Project: Specifying Legal Terms of Contract in ICT Environment", Journal of ITCON, www.itcon.org/2001/12/.

Cooper, R., Wootton, A., Bruce, M., Morris, B., Roberts, A., (1998), "A Generic Guide to Requirements Capture", The Research Centre for Design and Manufacture, The University of Salford, Greater Manchester, United Kingdom.

Eberlein, A., Leite, J.C., (2002), "Agile requirements definition: a view from requirements engineering", Time Constraint Requirements Engineering Workshop, IEEE Joint International Requirements Engineering Conference, Essen, Germany, http://www.enel.ucalgary.ca/tcre02/.

Faraj, I., Alshawi, M., Aouad, G., Child, T., Underwood, J., (2000), "An IFC web-based collaborative construction computer environment: WISPER", Proceedings of the National Conference on Objects and Integration for AEC, UK, ISBN 186081 3771.

Froese, T., Paulson, B.C. Jr., (1994), "OPIS: An Object Model-Based Project Information System", Microcomputers in Civil Engineering, 9(1):13–28.

Greening, R., Edwards, M., (1995), "ATLAS Implementation Scenario". In R. Scherer, editor, Proceedings ECPPM'94: Product and Process Modelling in the Building Industry, Dresden, Germany, pp. 467–472.

Hammersley, M., Atkinson, P., (1983), "Ethnography, Principles and Practice", London, Tavistock Publications.

Keil, M., Cule, P.E., Lyytinen, K., Schmidt, R.C., (1998), "A Framework for Identifying the Software Projects Risks", Communication of the ACM, 41:76–83.

Lavrac, N., Ljubic, P., Jermol, M., Papa, G., (2005), "A Decision Support Approach to Modelling Trust in Networked Organisations", Book chapter in Innovations in Applied Artificial Intelligence, Lecture Notes in Computer Science, Berlin, Heidelberg, Springer, ISBN 978-3-54026551-1.

LEGAL-IST, (2005), "Legal Issues for the Advancement of the Information Society Technologies – Project", www.legal-ist.org.

Maguire, M., (1996), "Prototyping and Evaluation Guide", Leicestershire, HUSAT Research Institute.

McPhee, C., Eberlein, A., (2002), "Requirements engineering for time-to-market projects", Proceedings of the 9th Annual IEEE International Conference on the Engineering of Computer Based Systems ECBS2002, Lund, Sweden.

Sarshar, M., Betts, M., Abbott, C., Aouad, G., (2000), "A Vision for Construction IT 2005–2010", RICS (Royal Institute of Chartered Surveyors), Research Series, pp. 1–42.

Sarshar, M., Tanyer, M., Aouad, G., Underwood, J., (2002), "A Vision for Construction IT 2005–2010: Two Case Studies", Engineering, Construction and Architectural Management, 9(2):152–160.

Scherer, R., editor, (1995), "EU-project COMBI–objectives and overview, proceedings", ECPPM'94: Product and Process Modelling in the Building Industry, Dresden, Germany, Balkema, pp. 503–510.

Sommerville, I., Sawyer, P., (1997), "Requirements Engineering: A Good Practice Guide", Chichester, England, John Wiley and Sons.

Sun, M., Aouad, G., (2000), "Integration Technologies to Support Organisational Changes in the Construction Industry", 7th ISPE International Conference on Concurrent Engineering, Lyon, France, 596–604.

Sun, M., Aouad, G., Bakis, N., Birchall, S., Swan, W., (2000), "GALLICON, A Prototype for the Design of Water Treatment Plants Using an Integrated Project Database", International Journal of Computer Integrated Design and Construction, 2(3):157–165.

Sun Microsystems, Inc., (2004), "Identity Management: Technology Cornerstone of the Virtual Enterprise", http://www.sun.com/software/products/identity/wp_virtual_enterprise.pdf, Retrieved March 7, 2008.

Wiegers, K.E., (2003), "Software Requirements 2: Practical Techniques for Gathering and Managing Requirements Throughout the Product Development Cycle", 2nd edition, Redmond, Washington, Microsoft Press, ISBN 0-7356-1879-8.

2

Requirements Engineering in Software Development

2.1 Introduction

Modern industrial information systems are complex high-technology systems made up of multiple interacting subsystems. There are a variety of business systems, instrumentation systems, information systems, infrastructure systems and management systems that interact to make the overall system work. The process of information system development has been seriously affected by the rapid development of technologies. In the past, it was possible for designers to understand and predict the components that would be available when the system was finally deployed and installed at the end user's work environment. Follow-up support and assistance was less complicated then than it is today. In order to overcome today's challenges, uncertainties and ambiguities, the development processes should be both rapid and flexible. The creation and management of large-scale complex systems have resulted in the emergence of management and engineering functions. Technical management is essentially an engineering discipline and it is expressed as systems engineering (Carr, 2000).

The practice of systems engineering embraces many of the critical activities required to deliver a new system. All of these activities are valuable but the most valuable function of systems engineering is formal requirements engineering (RE) and management. It is widely acknowledged within the software industry that projects are critically vulnerable when these activities are performed poorly (Abran *et al.*, 2004). The term *requirements engineering* is widely used in the field to denote the systematic handling of requirements because it directly affects the success or failure of the system to be deployed. Supply side issues have dominated techniques, processes and methods of software development for the past 40 years; this has resulted in technology-oriented developments rather than user-oriented ones. To achieve the levels of functionality, flexibility and successful deployment, RE is required to have a more demand-centric view in system development (Bennett *et al.*, 2000). Some benefits of good requirements include the following:

- Agreement among developers, customers and users on the job to be done and the acceptance criteria for the delivered system
- A sound basis for resource estimation (cost, personnel quantity and skills, equipment and time)
- Improved system usability, maintainability and other quality attributes
- Achievement of goals with minimum resources (less rework, fewer omissions and misunderstandings).

RE is the field of software requirements knowledge that is concerned with the elicitation, analysis, specification and validation of software requirements. This software requirements knowledge area is closely related to the software design, testing, maintenance, configuration management, engineering management, engineering process and quality knowledge areas. Figure 2.1 shows the breakup of the subjects covered by the software requirements knowledge area.

2.2 Requirements Engineering

At its most basic, RE is concerned with the real-world problems to be addressed by a software system. The user tasks and functions, business processes and devices are typically complex. By an extension, therefore, the requirements of particular kinds of software are typically a complex combination of requirements from different people at different levels of an organisation and from the environment in which the software will operate (Abran *et al.*, 2004). Therefore, RE is an iterative process by which the needs, preferences and requirements of individuals and groups who are significant to the development are researched, identified, elicited and modelled. The following thus should at least be defined (Cooper *et al.*, 1999):

- Customer, user and market requirements
- Design requirements
- Technical requirements.

RE does not include a stand-alone process. In order for RE to operate effectively, there are at least two other aspects, which must be in place (Cooper *et al.*, 1999):

- A clear business strategy – both long term and short term
- A formal product development process.

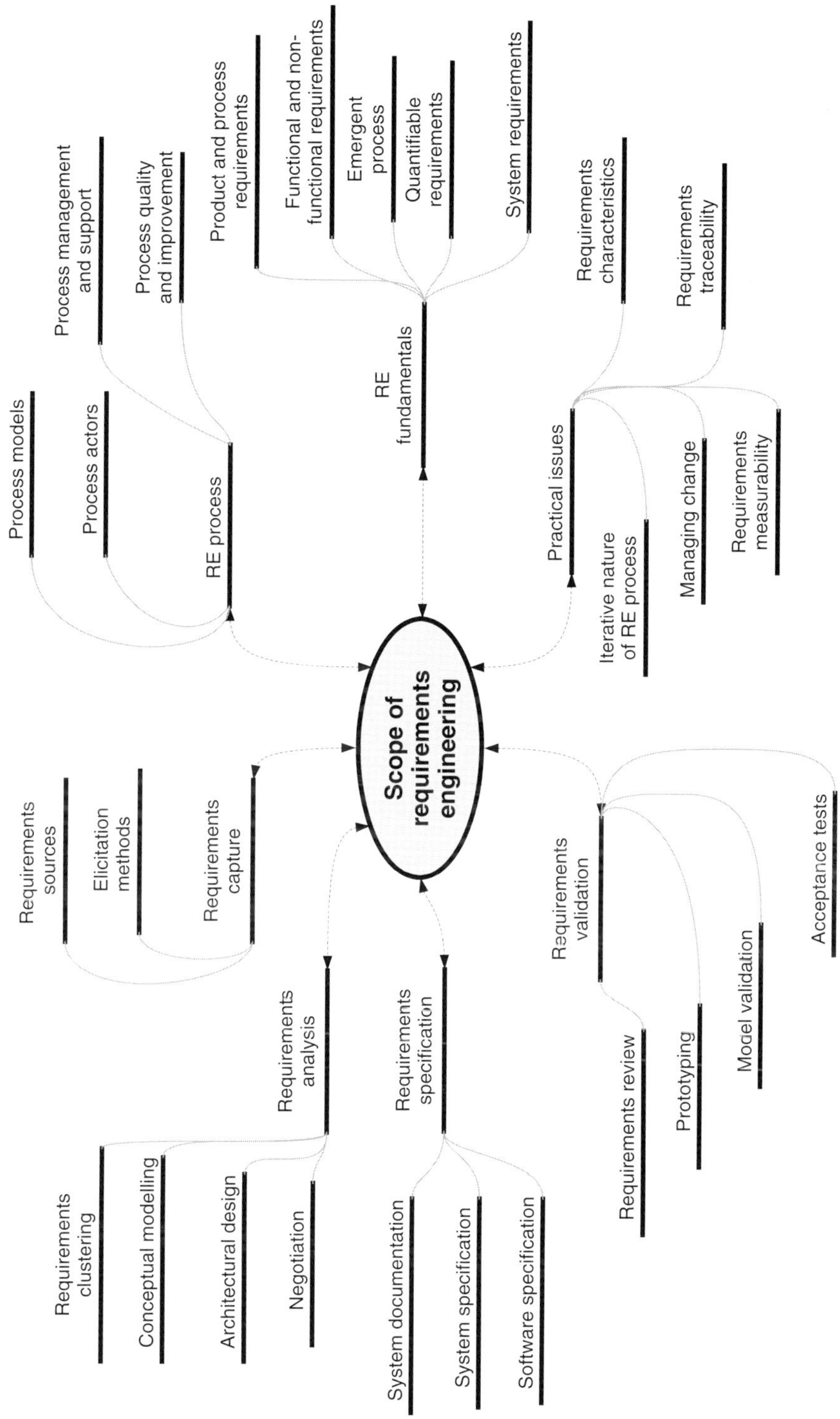

Figure 2.1 The requirements engineering knowledge coverage (based on Wiegers, 2003).

Business strategy both informs and is informed by RE. At the beginning of the development life cycle, business strategy provides the criteria by which all ideas and concepts are screened. The process of RE needs a business strategy in order to focus its activities, such as selecting information sources, prioritising requirements, and screening ideas and concepts. On the other hand, a business strategy needs RE in order to gather detailed information about evolving industrial and technological environments. They are interdependent; one cannot function effectively without the other.

In order for RE to be truly effective, it needs to form a part of a formal development process. Such a development process provides a road map from concept to deployment and beyond deployment.

Capturing requirements seems to be a simple matter that can be done quickly and put aside. But studies, surveys and experience demonstrate that a poorly executed RE is the biggest contributor to system failures, such as late delivery, excessive budget and either poor or incorrect performance (Carr, 2000). Superficially conducted RE causes major risks to a software development project (Nikula and Sajaniemi, 2002).

For example, the software projects risk identification framework by Keil *et al.* (1998) includes 11 risks factors and 6 of them are related to requirements. Furthermore, the CHAOS (1995, 2000) survey (The Standish Group, 1995 and 2001) identified the 14 factors that can cause the failure in software development; 6 of these factors are related to user requirements risk factors, only one of which is not mentioned in the risk identification framework (Keil *et al.*, 1998). Furthermore, Carr (2000) also addressed the RE-related risk factors that can also cause project failure. The combined set of RE-related risks factors contains the following issues:

1. Misunderstanding and misinterpretation of the requirements
2. Lack of adequate user involvement
3. Failure to manage end user expectations
4. Changing scope and objections
5. Lack of frozen requirements
6. Conflict between user departments
7. Incomplete and inconsistent requirements and specifications
8. Ambiguous and vague requirements.

RE as depicted in Figure 2.1 should be performed throughout the project life cycle, but it is most effective earlier in the process. There is a misleading perception that RE activities are produced only on paper, which does not seem to be real progress until the system is actually under construction.

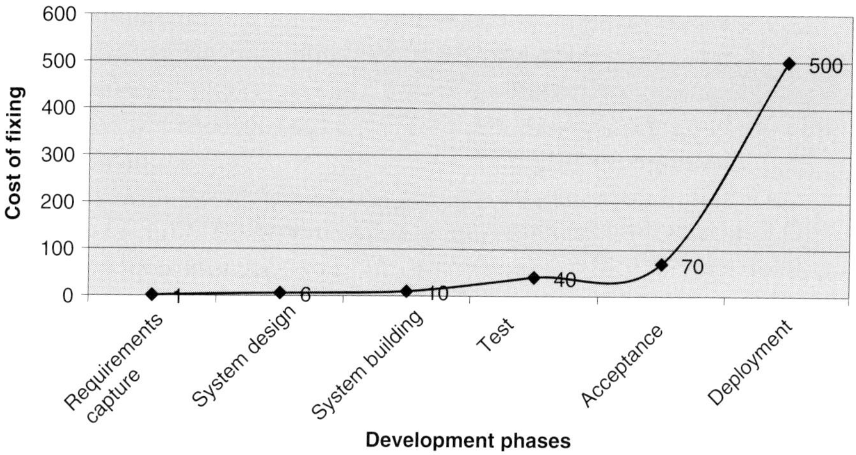

Figure 2.2 Cost ratio of fixing a problem in systems design (based on Carr, 2000).

It sometimes seems that programmers are willing to spend huge amounts of time and money to fix things later, rather than doing it right the first time. But the truth is simple: spending time and money in doing proper work in RE and management saves a large amount of time and money as illustrated in the Figure 2.2, which shows the relationship between the costs of fixing a problem according to the phase in which it was discovered. The figure implies that the later the problem is fixed, the higher the cost, with dramatic escalation of costs occurring after deployment. There is a general rule of thumb that the earlier an error occurs, and the later it is discovered, the more expensive it is to fix. Problems that arise due to poor RE are cheap to fix early in the development of the system. However, it is extremely expensive as and when the development cycle draws to an end.

For example, if a problem costs £1 to fix during requirements capture, it will cost £6 during design, £10 if the problem is found during the building phase, £40 if found during the development and testing phase, £70 if found during acceptance testing phase and £500 if found after deployment. On the other hand, studies about project success factors have also created interest and attention in order to identify RE practices that can be considered as software project success factors (Nikula, 2003).

The study by Hofmann and Lehner (2001) included 15 RE teams including 6 commercial off-the-shelf (COTS) and 9 customised application development projects. The project performances were measured in three areas: the quality of the RE service, the quality of RE products and process control. The study approached RE from an integrated viewpoint and investigated how team knowledge, allocated resources and deployed RE processes contributed to project success. As a result of this study, Hoffmann and Lehner (2001) proposed 10 best practices for successful RE to contribute to project success. These are as follows:

- Involve customers and users throughout RE for a better understanding of real needs.
- Identify and consult with all likely sources of requirements to improve the requirements coverage.
- Assign skilled project managers and team members to RE activities for more predictable performances.
- Allocate 15 to 30% of total project effort to RE activities to maintain the high quality specification throughout the project.
- Provide specification templates and examples to improve the quality of specification.
- Maintain good relationships among stakeholders to satisfy user needs better.
- Prioritise requirements to focus on the most important user needs.
- Develop complementary models together with prototypes to eliminate specification ambiguities and inconsistencies.
- Maintain a traceability matrix to establish explicit linkages between requirements and work products.
- Use peer reviews, scenarios and walk-throughs to validate and verify requirements for more accurate specifications and higher end user satisfaction.

In addition to the above, the requirements-related success factors in the CHAOS report are (i) user involvement, (ii) clear statement of requirements and (iii) realistic expectations and clear vision and objectives (Nikula, 2003).

2.3 Requirements Fundamentals and Principles

Elicitation, analysis and validation are at the heart of the RE process. Requirements elicitation is concerned with where software requirements come from and how they can be collected. It is the first stage in building an understanding of the problem that the software is required to solve. It is fundamentally a human activity, and this is where the stakeholders are identified and relationships established between the development team and the customer (Pfleeger, 2001), variously known as *requirements capture*, *requirements discovery* and *requirements acquisition*; however, they all refer to requirements elicitation.

One of the fundamental views of good RE is that there should be good communication between software users and software engineers. Before development begins, requirements specialists may begin the string of communication. They must mediate between the domain of the software users (and other stakeholders) and the technical world of the software engineers. Requirements analysis is about the following (Somerville, 2005):

- Detecting and resolving conflicts between requirements
- Discovering the bounds of the software and how it must interact with its environment
- Elaborating system requirements to derive software requirements.

The traditional view of requirements analysis is the conceptual modelling using some analysis methods such as the Structured Analysis and Design Technique or Contextual Design approach. Care and focus must be given to describe requirements precisely to enable the requirements' validation and their verification for implementation, as well as their accurate costs estimation. On the other hand, the requirements documents are subject to validation and verification procedures. The requirements should be validated to ensure that the software engineer has understood the requirements, and it is also important to verify that a requirements document conforms to company standards and that it is understandable, consistent and complete. Different stakeholders, including representatives of clients and the developer, should review the documents (Pinheiro, 2002).

It is normal to explicitly schedule one or more points in the requirements process where the requirements are validated. The aim is to pick up any problems before resources are committed to addressing the requirements. Requirements validation is concerned with the process of examining the requirements document to ensure that it defines the right software (Kotonya and Sommerville, 2000). These elicitations, analysis and validation activities are associated with the following requirements principles (Pinheiro, 2002):

- **Purposefulness:** There should be an objective to be fulfilled.
- **Appropriateness:** Requirements should be appropriate to the system. They should express what is necessary to achieve the system's objectives.
- **Truthfulness:** Requirements should express what is actually required.

The first two principles are closer to elicitation and analysis, and the third is closer to validation. These principles are also related to requirements properties such as correctness and consistency (Pinheiro, 2002). Roman (1985) uses the term *appropriateness* when referring to requirements specification. Nuseibeh and Easterbook (2000) stated that one of the most important goals of elicitation is to find out what problem needs to be solved. In addition, they addressed the validation as a very difficult problem in regards to truth and knowledge.

2.3.1 Purposefulness

Software requirements express part of what is necessary to achieve the system's objectives by means of software. The determination of a system's objectives is a difficult task, aggravated by situations in which they cannot be clearly stated. Usually there are several objectives satisfying different interests, and determining the right objectives is the hard task. To make matters worse, the system may not have clear or absolute goals. However, there is always a negotiated construction of the objectives and the system will always serve some purpose.

2.3.2 Appropriateness

Software is a single component of large systems, and thus it has to be appropriate to the system in which it is inserted. Determining the appropriateness of the software implies capturing the right requirements. If the objectives are not clear, the determination of the right requirements is even more difficult.

2.3.3 Truthfulness

In the development of complex systems, a lot of mistakes and errors occur. Requirements analysis and the process of verification and validation are two ways to deal with those mistakes and errors on requirements. Requirements analysis establishes the presence of desirable attributes such as consistency, precision and completeness, whereas verification and validation is a process of building confidence. There are several ways of verification and validation. For instance, inspection and walk-through sessions, scenario building, prototyping or incremental development and testing are some techniques for requirements validation. Testing is especially a necessary step, not only to identify errors introduced in the construction of various artefacts but also to build confidence in the truthfulness of the system.

2.4 Requirements Engineering Process

Software RE process is the process of determining what is to be produced in a software system development. In developing a complex software system, the RE process has the widely recognised goal of determining the needs for, and the intended external behaviour of, a system design. This process is regarded as one of the most important parts of building a software system: 'The hardest single part of building a software system is deciding what to build. No other part of the conceptual work is as difficult as establishing the detailed technical requirements, including all the interfaces to people, to machines, and to other software systems. No part of the work so cripples the resulting systems if done wrong. No other part is more difficult to rectify later' (Brooks, 1987).

The RE process is firstly not a discrete front-end activity of the software life cycle, but rather a process initiated at the beginning of a project and continues to be refined throughout the life cycle. Secondly, it identifies software requirements as configuration items and manages them using the same software configuration management practices as other products of the software life cycle processes and thirdly, it needs to be adapted to the organisation and project context.

In particular, it is concerned with how the activities of elicitation, analysis, specification, and validation are configured for different types of projects and constraints. It also includes activities that provide input into the requirements process, such as marketing and feasibility studies (Robertson and Robertson, 1999; Sommerville, 2005).

Tracing the emergence of significant ideas in software development over the years, one can observe that in the 1960s attention was on coding, in the 1970s on design and in the 1980s on specification. However, in the last two decades, RE has become a critical issue. The challenge is that requirements describe the 'what' of a system, and not the 'how'. However, in the process of RE it is often difficult to state the real 'what' level of a system because one person's 'how' may be another person's 'what' and the converse. In this perspective, the requirements engineer faces a complex problem in meeting the needs of the end users and at the same time meeting the needs of the designer.

There are a number of stakeholders who participate in the requirements process. This process is fundamentally interdisciplinary, and the requirements specialist needs to mediate between the domain of the stakeholders and that of software engineering. The stakeholders will vary across projects, but always include users or operators and clients.

Typical examples of software stakeholders include but are not limited to the following (Abran *et al.*, 2004):

- **Users:** This group comprises those who will operate the software. It is often a heterogeneous group comprising people with different roles and requirements.
- **Clients:** This group comprises those who have commissioned the software or who represent the software's target market.
- **Market analysts:** A mass-market product will not have commissioning clients, so marketing people are often needed to establish what the market needs and to act as proxy clients.
- **Regulators:** Many application domains such as banking and public transport are regulated. Software in these domains must comply with the requirements of the regulatory authorities.
- **Software engineers:** These individuals have a legitimate interest in profiting from developing the software.

It will not be possible to perfectly satisfy the requirements of every stakeholder, and it is the requirements engineers' job to negotiate trade-offs that are both acceptable to the principal stakeholders and within budgetary, technical, regulatory, and other constraints. A prerequisite for this is that all the stakeholders are identified, the nature of their stake analysed and their requirements elicited (You, 2001).

In addition to the requirements elicitation, analysis and validation activities, there is a fourth generic activity in the RE process, which is called *requirements management*. It concerns the project management resources required and consumed in the RE process. Its principal purpose is to make the link between the process activities (elicitation, analysis and validation) and the issues of cost, human resources, training and tools (You, 2001; Lundh and Sandberg, 2002).

Although these core activities seem to be separate tasks, these four activities cannot be strictly separated and performed sequentially. Some of the requirements are implicit in the working practices, while others may only arise when design solutions are proposed. All four activities are performed repeatedly because the needs are often impossible to realise until after a system is built. Even when requirements are stated initially, it is likely that they will change at least once during development and it is very likely they will change immediately after development.

There is a variety of techniques and methodologies to conduct the core RE activities such as requirements elicitation, requirements analysis and requirements validations. Figure 2.3 below shows some methods undertaken in discrete stages of the development life cycle for user-centred design.

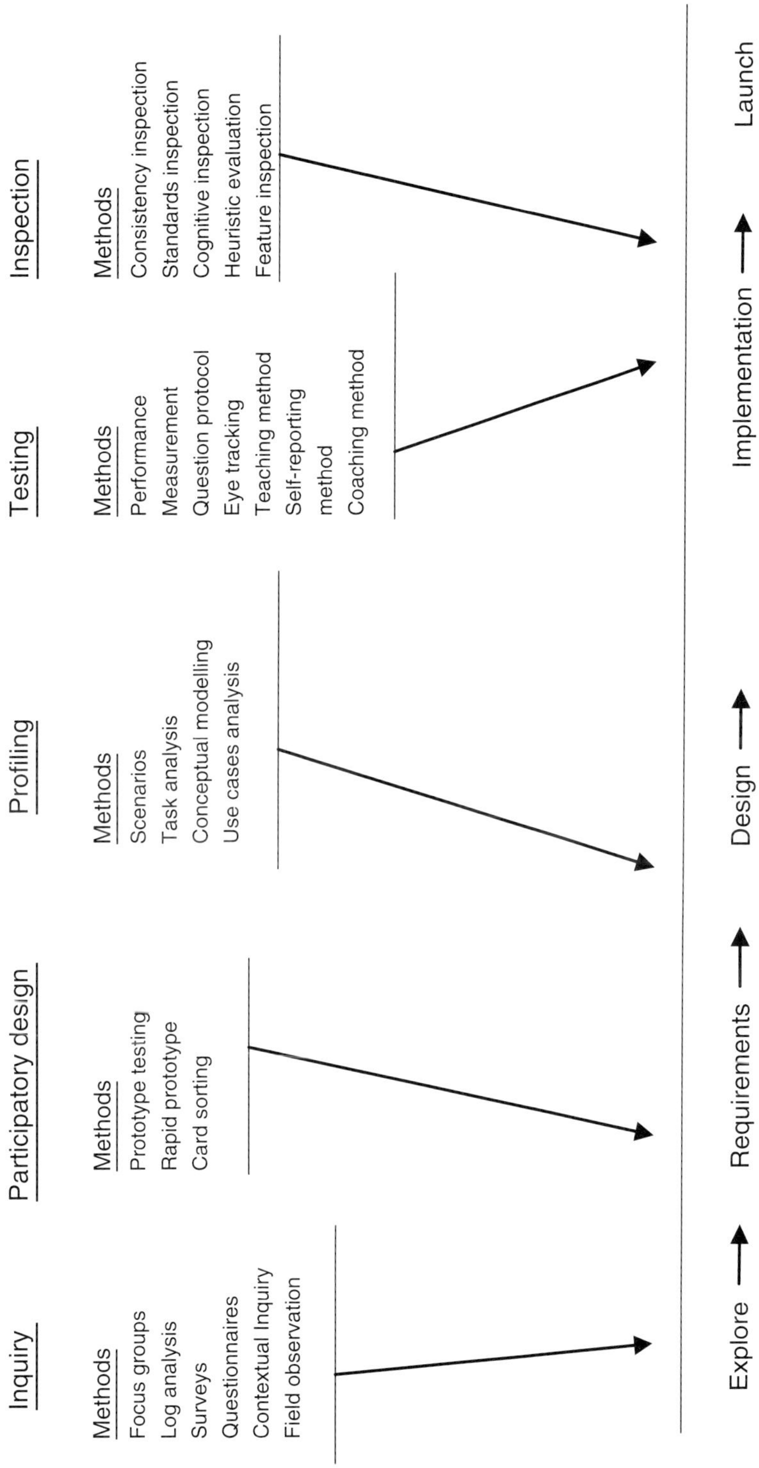

Figure 2.3 Mapping the common requirements methods on the development life cycle.

McPhee and Eberlein (2002), in their survey-based research, had a goal of gaining a greater insight into how software developers view the RE process and what tools and techniques they consider as being most useful for it. Another goal was to clarify the possible differences between time-to-market (TTM) projects and non-TTM projects in their approaches to RE. One part of their study addressed the level of knowledge and perceived usefulness of different RE tools and techniques such as scenario development, requirements prioritisation, change management, evolutionary prototyping, testing, checklisting and interviews.

McPhee and Eberlein (2002) investigated the reasons for choosing RE techniques and their desirable qualities through survey. For example, previous experience, enabling good communication, company standards and available tool supports for the targeted RE techniques are the primary reasons for choosing the relevant RE techniques. In terms of desirable quality of RE techniques, major justifications are easy to use, enabling the capture of the complete set of requirements, allowing traceability, improving product quality, tool support and helping in prioritisation. Another survey-based research conducted by Doherty and King (1998) discussed that organisational issues play a substantial or even primary role in systems failures. This confirms the increasing importance of organisational issues and the need to address them explicitly in a systems development process. These organisational issues are illustrated in Figure 2.4.

The effort of designing usable, targeted, desirable and efficient systems involves acquiring and analysing user data. Selecting exactly the right method for this can be difficult without a clear understanding of how each method is used, what its possible benefits are and how it will affect the project budget. Therefore, defining the right techniques and approaches are very important.

The techniques in Figure 2.3 can be used in the focused phases of the development life cycle. However, for innovative large-scale information systems such as the computer integrated environments (CIE) systems for the construction industry, requirements activities must be carried out from the outset up to the deployment. Therefore, there must be a coherent relationship and link between the requirements techniques throughout the system development life cycle. For instance, researchers and engineers acknowledge that the use case technique is a good tool for requirements elicitation and analysis but it is inadequate to capture the whole requirements of the system such as non-functional requirements (Castro *et al.*, 2001; Arayici *et al.*, 2006).

RE process has been acknowledged as a critical process of software development, precisely because it deals not only with technical

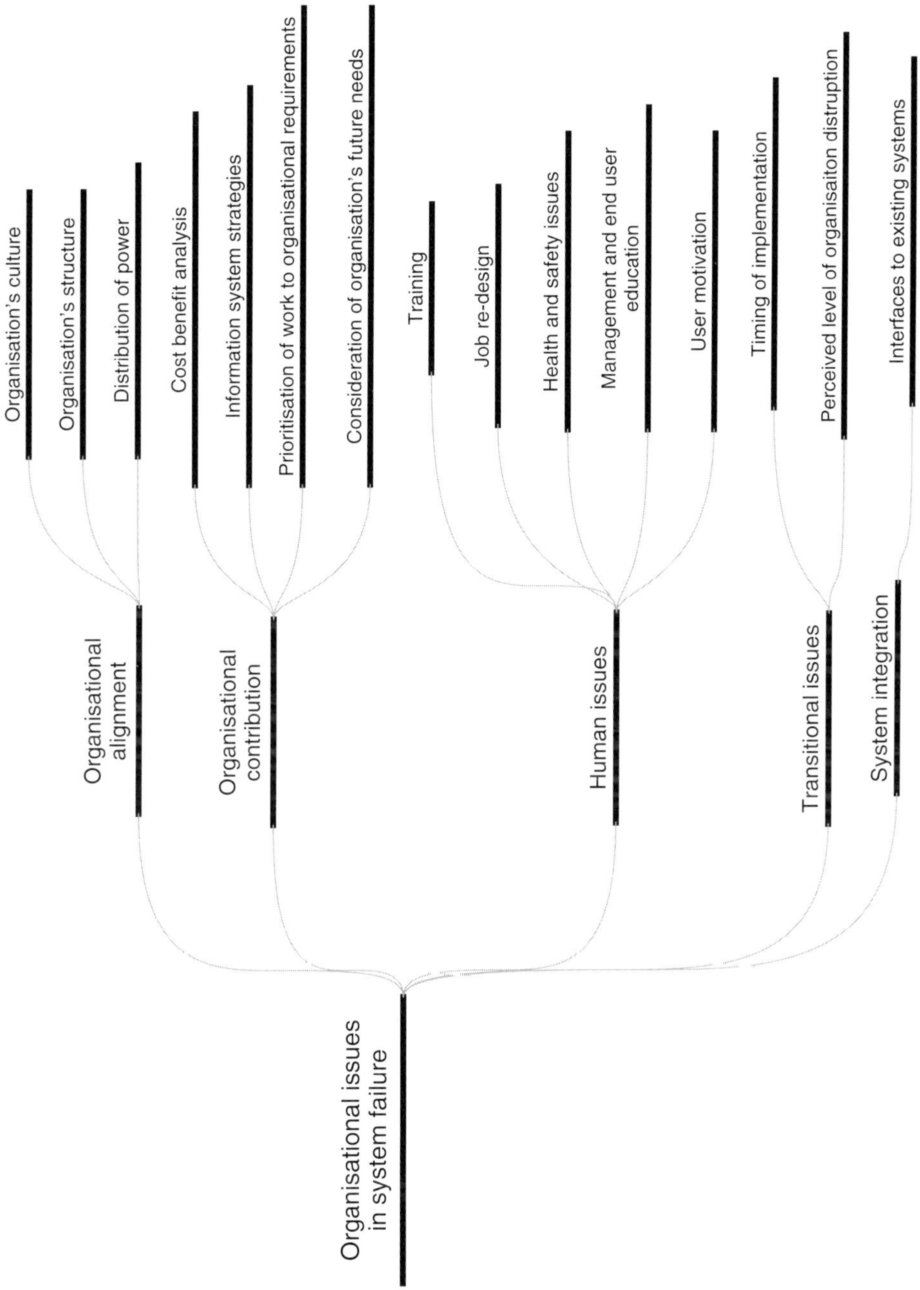

Figure 2.4 Organisational issues in system developments that can cause system failures (Doherty and King, 1998).

knowledge but also with organisational, managerial, economic and social issues.

There is emerging consensus that a requirement specification should include not only software specifications but also business models and other information describing the context within which the intended system will operate (Erikson and Penker, 2000). Consequently, there is a need for modelling and analysis of stakeholder interests and how they might be addressed, or compromised, by various system and environment alternative structures (Castro *et al.*, 2001).

The RE process has drawn an important distinction between the early design phase RE and the late-phase requirement engineering. The early-phase requirements capture activities are typically informal and address organisational or non-functional concerns. The emphasis during this phase is on understanding the motivation and rationale that underlie system requirements. Late-phase activities, on the other hand, focus on completeness, consistency and verification and validation of requirements.

The dominant object-oriented modelling technique and Unified Modelling Language (UML) (Booch *et al.*, 1999) are well suited only for the later stages of the RE process (Castro *et al.*, 2001). It is argued that UML is ill-equipped for early RE because it cannot represent how the intended system meets organisational goals, why the system is needed, what alternatives were considered, what the implication of the alternatives are for the various stakeholders and how the stakeholders' interests and concerns might be addressed.

Therefore, researchers have synthesised use cases with additional techniques by placing these techniques before or after the use case development and analysis. For example, use case models are actually a loose collection of use cases and these use case models in the use case-driven requirements analysis have some weaknesses for requirements elicitation and analysis. Regnell *et al.* (1995) identified the following weaknesses of use case-driven analysis (UCDA) in general:

- Use cases are not independent.
- Use cases occur under specific conditions. They have invocation and termination contexts.
- The level of abstraction of use cases and their length are matters of arbitrary choice.
- The use case can, in practice, guarantee only partial coverage of all possible system usage scenarios.

Therefore, many RE approaches improve the original UCDA, for example, by extending it with a synthesis phase where separate use cases are integrated into a synthesised usage model in which a strong and coherent relationship between the use cases is established after the use cases were initially developed (Regnell *et al.*, 1995). For example, in the contextual design approach by Beyer and Holtzblatt (1998) describes an RE methodology in which the use cases and object models of UML also exist. However, before the use case development starts, there are some priory requirements activities defined such as visioning, storyboarding and user environment design (UED). Beyer and Holtzblatt (1998) defined a coherent and strong tie between these subsequent activities.

Completion of the preliminary activities enables to capture non-functional requirements as well as developing more coherent and consistent use case and object models. As a result, Beyer and Holtzblatt proposed an RE approach, a so-called contextual design, for both requirements elicitation and analysis. Since RE is a dynamic subject that keeps changing throughout the development life cycle, the requirements needs is continuously being updated and the system under development should align with the requirements all the time. To establish this alignment between the requirements and the system under development, further RE techniques are required for the validation of the requirements as well as the validation of the system.

The agile processes such as incremental prototyping, testing, user involvement, etc., for the validation of user requirements are state-of-the-art RE techniques. Subsequently, these coherent relationships between the consecutive requirements approaches help to effectively conduct the requirements elicitation, analysis and validation, respectively.

Requirements management is also facilitated throughout the project life cycle by establishing a coherent relationship and synthesis between the RE activities. A foundation of the coherent synthesis between the requirement methods has resulted in RE processes occurring throughout the life cycle. In the following section, the RE approaches address that the issues in requirements elicitation, analysis, and validation are explained in detail.

2.4.1 Contextual design approach

Contextual design is an approach to define software and hardware systems, which collects multiple user-centred techniques into an integrated design process. The methodology challenges how that data is gathered from the end users, the base criteria for deciding what the system should do, and how it should be structured. It helps decide on

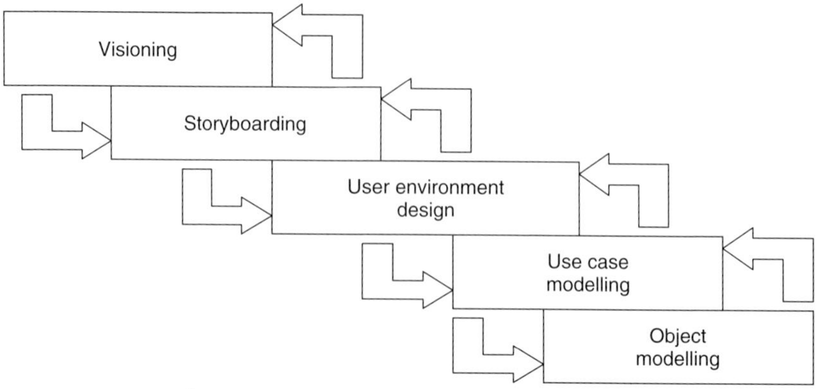

Figure 2.5 The contextual design approach (based on Beyer and Holtzblatt, 1998).

how the end users will work in the future and assesses the core design problem and uses those decisions to drive the use of technology. It unifies all the actions of an organisation into a coherent response to the end users and their work environment, rather than leaving team members to argue with each other on the basis of personal opinion, anecdotes or unverifiable claims about 'what the end users would like' (Beyer and Holtzblatt, 1998). The methodology, which is depicted in Figure 2.5, has five main stages. These are visioning, storyboarding, user environment design, use case modelling and object modelling.

Vision development

A vision captures the scope as a single picture, showing all parts of the vision together. It says what the new work practice will be without showing how it will happen over time. Drawing a vision follows the following steps.

A. **Contextual Inquiry:** Contextual inquiry uncovers who the end users really are and how they work on a day-to-day basis. The cross-functional design team conducts one-on-one field interviews with the end users in their workplace to discover what matters in the work. Team members observe people as they work and inquire into actions as they unfold to understand their motivations and strategy. The interviewer and the end users, through discussion, develop a shared interpretation of the work. Within contextual inquiry, team interpretation sessions are established, bringing the design team together to hear the whole story of each interview and capture the insights and learning relevant to their design problem.

An interpretation session brings the unique perspectives of all team members to the data, sharing design, marketing and business implications. Through these discussions, the team captures issues, draws work models and develops a shared view of the end users whose data and needs are being interpreted. As an output of the team interpretation sessions, work models are developed in the vision stage.

B. **Work Modelling:** Work models are produced as a result of contextual inquiry through the team interpretation sessions. Work models capture the work of individuals and organisations in diagrams. Five models provide different perspectives on how work is done:
   1. The flow model captures communication and coordination.
   2. The cultural model captures culture, policy and influences.
   3. The sequence model shows the detailed steps performed to accomplish a task.
   4. The physical model shows the physical environment as it supports the work.
   5. The artefact model shows how artefacts are used and structured in doing the work.

C. **Consolidation:** After work models are developed, consolidation needs to be done within the vision development. It brings data from individual user interviews together to build up a shared understanding so that the team can see common patterns and structure without losing individual variation. In addition, the affinity diagram brings together issues and insights across all users into a wall-sized, hierarchical diagram to reveal the scope of the problem (Beyer and Holtzblatt, 1998).

   Consolidated work models bring together different types of work models in order to reveal common strategies and intents while retaining and organising individual differences. Together with the affinity diagram, consolidated work models produce a single rich picture, which represents the vision for development.

Storyboard

Once a shared vision is developed by means of work modelling and affinity diagrams, the details of the vision should be transformed into storyboards that show how specific tasks will be accomplished in the new vision. A storyboard demonstrates the new procedure for doing a task pictorially. Each frame in the storyboard shows a single scene.

Re-engineering the work via storyboarding employs the consolidated work models and affinity diagrams to drive the process re-engineering in relation to how to improve work by using technology to support the work practice. This focuses on how technology helps people to fulfil their tasks, as opposed to focusing on what could be done with technology without considering the impact on the real lives of people. By means of storyboarding, a story of how the end users will do their work in the new world invented in the vision development is produced as the new process. The storyboard developed includes the system, its delivery and support structures to make the new work practice successful.

User environment design

The new system must have the appropriate function and structure to support the workflow described in the storyboard. The UED represents the structural floor plan of the new system. It shows how each part of the system supports the user's work, exactly what function is available in that part and how the end user may/would have access to and from other parts of the system.

Paper prototyping is conducted in this stage to iteratively improve the structure design of the system together with the end users. Testing is an important part of any systems development. It is generally accepted that the sooner problems are found, the less it costs to fix them. So it is important to test and iterate a design early, before anyone invests in the design and before spending time writing the code. The simpler the testing process, the more is the iteration conducted to work out the detailed design with the users.

Paper prototyping develops rough mock-ups of the system using post-its to represent windows, dialog boxes, buttons and menus. The design team tests these prototypes with the end users in their workplace by replaying real work events in the proposed system. When the user discovers problems, they and the designers redesign the prototype together to fit their needs.

Refining the design with users gives designers a user-centred way to resolve disagreements and work out the next layer of the requirements. Several paper prototype sessions are conducted to improve the system and to complete a detailed user interface design.

Use cases and object-oriented design

In object-oriented design via use cases, object modelling does not stand on its own. The purpose of a use case is to tell a coherent story of how the end users will work and the system will meet their needs (Jacobson *et al.*, 1995). Use cases and object models have a user-centred foundation from the storyboarding and the UED stages.

Prioritisation helps the transition to implementation by planning the system implementation over time. A well-defined process walks the development team through a design and breaks it into coherent delivery units, each of which delivers value to the end users (Beyer and Holtzblatt, 1998). Furthermore, object-oriented design helps to move from systems design to the design of the implementation. The development team walks through the development of use cases based on the UED and storyboards, and then translates these use cases into an object design.

2.4.2 Use case-driven requirements analysis

In UML, requirement is defined as a condition or capability to which a system must conform, and requirements are analysed in two groups: functional requirements and non-functional requirements. Functional requirements are used to express the behaviour of a system by specifying both the input and output conditions that are expected to result. Non-functional requirements are used to express a wide variety of attributes that are described specifically by the system's non-functional requirements such as usability, reliability, performance and supportability (Kruchten, 2000). In the rational unified process, the requirements workflow is as shown in Table 2.1 (Kruchten, 2000).

In the UML domain, the UCDA plays an important role in RE. The basic concepts of the UCDA are actors and use cases. *An actor* is a specific role played by a system user, and represents a category of users that demonstrate similar behaviour when using the system. *A use case* is a system usage scenario for specific actors. During the analysis, a number of typical use cases for every actor is described and identified. Use cases are expressed in natural language with the terms from the problem domain. UCDA helps to cope with the complexity of the requirements analysis process. Identifying, and then independently analysing different use cases enables the developers to focus on one narrow aspect of the system usage at a time.

Since the idea of UCDA is simple and the use case descriptions are based on natural concepts that can be found in the problem domain, the end users can actively participate in requirements analysis. Consequently, the developers can learn more about the potential users, their actual needs, and their typical behaviour. On the other hand, the lack of synthesis is the main drawback of UCDA. Use case model is

Table 2.1 Requirements workflow in rational unified process (based on Kruchten, 2000).

	Stages	Description
1	Analysing the problem	Gaining agreement on a statement of the problem to be solved, identifying stakeholders and identifying the boundaries and constraints of the system
2	Understanding stakeholders' needs	Using various elicitation techniques, gathering stakeholders' requests and obtaining a clear understanding of the real needs of the users and stakeholders of the system
3	Defining the system	Based on input from the stakeholders, establishing the set of the system features to be considered for delivery; determining the criteria that will be used to prioritise them and beginning to set realistic expectations with the stakeholders on what features will be delivered; identifying actors and use cases needed for each key feature
4	Managing the scope of the system	Collecting important information from the stakeholders in the project and maintaining those – along with the requirements – as requirement attributes to be used in prioritising and scoping the agreed set of requirements, so that the system can be delivered on time and on budget to meet the users' need expectations
5	Refining the system definition	Using a use case model, detailing the software requirements of the system to come to an agreement with the client on the functionality of the system and capturing other important requirements, such as non-functional requirements and design constraints, etc.
6	Managing changing requirements	Using requirements attributes and traceability to assess the impact of changing requirements, using a central control authority such as Change Control Board to control changes to the requirements, maintaining an agreement with the client and setting realistic expectations on what will be delivered

just a loose collection of use cases. And they can be very high level or very low level use cases. That is to say, use cases are useful but they are not scaled without any prior RE techniques.

In the subsequent phases of objectory, these use cases are directly used to create the analysis model. This model describes the structure of the system and is a step towards design. The main output of the requirement analysis is a model that captures the functional requirements and system usage, without any design aspects.

2.4.3 Agile requirements engineering processes

Building advanced and sophisticated systems is recognised as a complicated task by Schwaber (2002). When requirements are compared with technology, the degree of complexity increases as the requirements are less known and the technologies are less certain (Stacey, 2000). Frequent inspection and immediate adaptation to the results are the primary mechanism to deal with project complexity. Various implementations of the key practices shown in Table 2.2 support these mechanisms (Schwaber, 2002).

Table 2.2 Various implementation of agile techniques for tackling complexities in the development of CIEs.

	Key practices	Implementation
1	Iterative development	Frequent iterations generate increments that can be inspected to determine the state of the software development project and serve as a basis for adaptation
2	Increments of work	Composed of working system functionality rather than artefacts. These increments create a 1:1 relationship between progress and product delivery and provide a mechanism for user feedback to real product rather than hidden internal artefacts
3	Collaboration	The users and engineers form teams that work together
4	Daily meetings	Provide a daily picture of the internal status of the project
5	Adaptation	The team of developers arrange daily meetings, and the developers and the end users organise meetings at the end of every increment to guide the project in order to create the greatest value
6	Emergence	The architecture, team structure and requirements emerge during the course of the project rather than being determined at the outset. The team is guided by preliminary and sketchy visions of requirements and architecture. The architecture is initially elaborated in greater detail for large and complex systems. This simplifies the coordination of multiple teams

The cost of the project is expected to provide some business benefit or value that is greater than the cost of the project, which is unlikely in some projects because of poor performance of RE. In other words, RE is premised on the expectation that this value for return on business benefit can be accurately modelled in the requirements at the start of the project.

In the agile developments, the requirements enable the expected value to emerge. Emergence is used with increments of working functionality. Most users only respond to the system reality rather than the hidden internal artefacts as requirements are documented and modelled. When business conditions change and are often very different at any point in a project from the beginning to the end, agile processes allow user requirements to emerge dynamically without penalty to the project.

Agile processes implement the assertion through frequent collaboration between the end users and development teams to determine which requirements are the most valuable to develop next in a system under development. Eberlein and Leite (2002) and Pinherio (2002) addressed some important practices of RE that enrich agile processes. These are user involvement, analysis (verification and validation), non-functional requirements, managing change, incremental development and short delivery time paths. A brief explanation of the features of the agile RE practices is given below.

User involvement

The importance of user involvement in system development has already been recognised. Reviews and feedback from practitioners show that direct and regular interaction with the end users is one of the key factors for project success. Projects that fail tend to have a great lack of early end user feedback. It is important to identify all the individuals and organisations that will affect the success and failure of the system; after (or) following this identification of the stakeholders, requirements have to be elicited by means interviews, contextual inquiry, etc.

Analysis (verification and validation)

Agile methods rely fundamentally on the use of validation (testing), not only as a way of achieving quality by debugging errors but also as a way of identifying requirements through incremental prototyping.

The agile methods can benefit if verification is also taken into account together with validation by using simple tools to check early requirements descriptions associated with effective management practices. By doing the inspection for verification, the quality of the agile processes is improved. For example, user stories or test cases generated from user stories can be used for verification in the agile process.

Non-functional requirements

It is also common that failures are due to the lack of consideration of non-functional requirements. Considering non-functional requirements at the design stage causes major problems. However, if the non-functional requirements are taken into account at the implementation level, the problems will be even greater.

It is rare that the users address the specific transaction time, ease of use, levels of confidentiality, etc. Cysneiros (2002) reported that this type of difficulty in the realm of health-care applications and others have experienced similar difficulties when eliciting non-functional requirements from end users and clients in other domains (Eberlein and Leite 2002). Beck (2000) recommended that user stories that are scoped out by the end users should be utilised for the compilation of the non-functional requirements.

User stories

User stories are used to describe aspects to elaborate time estimates for planning each release and to develop the tests. They are written by the end users and express what the system needs to do for the end users. The stories should provide sufficient detail to make a possible determination of the time that will be required before implementation. During the actual implementation, the end users and developers test the software and the users provide further detailed descriptions of the requirements for further enhancement of the system.

Managing change

One of the most important aspects of agile methods is to manage the change in the project life cycle. Requirements management is the precondition of managing change based on the users' expressions. Central to requirements management is the concept of requirements traceability (Silva, 2002). For example, Extreme Programming (XP) is an agile method that employs user stories to analyse the user requirements for verification and validation. Breitman and Leite (2002) cited that by annexing the user stories as they correspond to the piece of code, the following will be gained:

1. The storage of the information in a persistent and retrievable manner since the user stories are the requirements
2. Traceability of the stored information by as date, risk, and priority
3. The possibility of deriving test cases from the user stories.

Incremental development

The use of prototypes naturally leads to a development based on successive prototypes. In other words, incremental prototyping is a scheduling and staging strategy in which various parts of the system are developed at different times and integrated on completion. The idea is to build the parts of the systems in increments. The development continues until the whole system is delivered.

The important feature in this form of development is the early delivery of useful functionalities. This is a great advantage and is very valuable in many circumstances including simplification of analysis tasks in each increment. The user requirements must be understood well enough so that the overall functionality can be clustered and the interface of each cluster well defined (Pinherio, 2002). Each increment is built to deliver the associated functionality and it should be prepared to work together with the next increments. A poorly understood system leads to a large amount of remedying with several problems, which will diminish any benefit gained from the functionalities delivered earlier.

The path to short delivery times

In order to clarify obscure issues in successive prototypes in the development life cycle, the increments should be based on a well thought out architecture and the user involvement should be carefully prepared and conducted. A short development time is a consequence of doing what should be done, in a better way.

Black-box tests

In each of the phases of iterative incremental development, the user stories are translated into test cases and included into a set of regression tests. At the end of the iteration, not only is the increment delivered but the entire release is also tested. Subsequently, all the bugs and the low functional and non-functional performances that were observed and analysed in the previous iterations can also be retested. Whenever an error is found during test operation, the error is scheduled for correction in future releases.

References

Abran, A., Séguin, N., Bourque, P., *et al.* (2004), "The Search for Software Engineering Principles – An Overview of Results", Buenos Aires, Argentina, Principles of Software Engineering.

Arayici, Y., Ahmed, V., Aouad, G., (2006), "A Requirements Engineering Framework for Integrated Systems Development for the Construction Industry", Electronic Journal of Information Technology in Construction, 11:35–55, www.itcon.org/2006/3.

Beck, K., (2000), "Extreme Programming, Explained Embrace Change", Massachusetts, Addison Wesley, Longman, Inc.

Bennett, K.H., Layzell, P.J., Budgen, D., Brereton, O.P., Macaulay, L., Munro, M., (2000), "Service-Based Software: The Future for Flexible Software", IEEE APSEC2000, The Asia-Pacific Software Engineering Conference, 5–8 December 2000, Singapore, IEEE Computer Society Press.

Beyer, H., Holtzblatt, K., (1998), "Contextual Design. Defining Customer-Centered Systems", San Francisco, California, Morgan Kaufmann Publishers.

Booch, G., Jacobson, I., Rumbaugh, J., (1999), "Unified Modelling Language User Guide", "Addison-Wesley Object Technology Series", Massachusetts, Rational Software Corporation.

Breitman, K.K., Leite, J.C.S.P., (2002), "Managing User Stories", IEEE Joint International Requirements Engineering Conference, Essen, Germany.

Brooks, F.P., (1987), "Essence and Accidents of Software Engineering", IEEE Computer, 20(4):10–19.

Carr, J., (2000), "Requirements Engineering and Management: The Key to Designing Quality Complex Systems", The TQM Magazine, 12(6):400–407; MCB University Press, ISSN 0954-478X.

Castro, J., Alencar, F., Filho, G., Mylopoulos, J., (2001), "Integrating organizational requirements and object oriented modeling", IEEE International Conference on Requirements Engineering, Toronto, Canada.

CHAOS Report, (1995), "Software Development Report", Boston, Massachusetts, The Standish Group, www.standishgroup.com.

CHAOS Report, (2000), "The Software Development Report", Boston, Massachusetts, The Standish Group, www.standishgroup.com.

Cooper, R., Bruce, M., Wootten, A., (1999), "Requirements Capture as a Process of Technology-Market Integration", International Journal of Technology Management, 17(6):582–596.

Cysneiros, L.M., (2002), "Requirements Engineering in Health Care Domain", In Proceedings of the RE02 International Conference on Requirements Engineering, Essen, Germany, IEEE Computer Society Press.

Doherty, N., King, M., (1998), "The Consideration of Organizational Issues During the System Development Process: An Empirical Analysis", Behaviour and Information Technology, 17(1):41–51.

Eberlein, A., Leite, J.C., (2002), "Agile Requirements Definition: A View from Requirements Engineering", Time Constraint Requirements Engineering Workshop, IEEE Joint International Requirements Engineering Conference, Essen, Germany, http://www.enel.ucalgary.ca/tcre02/.

Erikson, H., Penker, M., (2000), "Business Modelling with UML: Business Patterns at Work", New York, OMG Press, John Wileys & Sons.

Hoffmann, H., Lehner, F., (2001), "Requirements Engineering as a Process Factor in Software Projects", IEEE Software, 18(4):58–66.

Jacobson, I., Ericsson, M., Jacobson, A., (1995), "The Object Advantage Business Process Reengineering with Object Technology", Massachusetts, United States of America, Addison-Wesley Publishing.

Keil, M., Cule, P.E., Lyytinen, K., Schmidt, R.C., (1998), "A Framework for Identifying the Software Projects Risks" Communication of the ACM, 41:76–83.

Kotonya, G., Sommerville, I., (2000), "Requirements Engineering: Processes and Techniques", Massachusetts, United States of America, John Wiley & Sons.

Kruchten, P., (2000), "The Rational Unified Process – An Introduction", 2nd edition, Massachusetts, United States of America, Addison-Wesley, ISBN 0-201-70710-1.

Lundh, E., Sandberg, M., (2002), "Time constraint requirements engineering with extreme programming – an experience report", IEEE Joint International Requirements Engineering Conference, Essen, Germany.

McPhee, C., Eberlein, A., (2002), "Requirements engineering for time-to-market projects", Proceedings of the 9th Annual IEEE International Conference on the Engineering of Computer Based Systems ECBS2002, Lund, Sweden.

Nikula, U., Sajaniemi, J., (2002), "A lightweight combination of proven RE techniques", IEEE Joint International Requirements Engineering Conference, Essen, Germany.

Nikula, U., (2003), "Experiences from Lightweight RE Method Evaluations", IEEE International Workshop of Comparative Evaluation in Requirements Engineering, Monterey Bay, California, IEEE Computer Society, pp. 53–60.

Nuseibeh, B., Easterbrook, S., (2000), "Requirements Engineering: A Roadmap". In A.C.W. Finkelstein, editor, The Future of Software Engineering, Limerick, Ireland, IEEE Computer Society Press.

Pfleeger, S.L., (2001), "Software Engineering: Theory and Practice", 2nd edition, Prentice Hall.

Pinheiro, F.A.C., (2002), "Requirements honesty", IEEE Joint International Requirements Engineering Conference, Essen, Germany.

Regnell, B., Kimbler, K., Wesslen, A., (1995), "Improving the use case driven approach to requirements engineering", Proceedings of Second IEEE International Symposium on requirements Engineering, York, United Kingdom.

Robertson, S., Robertson, J., (1999), "Mastering Requirements Engineering Process", Guilford, Surrey, ACM Press, Addison Wesley, ISBN 0201360462.

Roman, G.-C., (1985), "A Taxonomy of Current Issues in Requirements Engineering", IEEE Computer, 18(4):14–22.

Schwaber, K., (2002), "The impact of agile processes on requirements engineering", IEEE Joint International Requirements Engineering Conference, Essen, Germany.

Silva, A., (2002), "Requirements, Domain and Specifications: A Viewpoint-Based Approach to Requirements Engineering", Proceedings of ICSE (IEEE International Conference on Software Engineering), Orlando, Florida, IEEE Computer Society Press.

Sommerville, I., (2005), "Software Engineering", 7th edition, Massachusetts, United States of America, Addison-Wesley.

Stacey, R.D., (2000), "Strategic Management and Organizational Dynamics, The Challenge of Complexity", 3rd edition, New York, Prentice Hall.

Wiegers, K.E., (2003), "Software Requirements", 2nd edition, Microsoft Press.

You, R.R., (2001), "Effective Requirements Practices", Boston, Massachusetts, United States of America, Addison-Wesley.

3

Computer Integrated Environments

3.1 Introduction

Corporations today are becoming largely dispersed and as a result are intensely founded on networking technology, allowing sharing and access to information in different locations. Meanwhile, computer-based information systems have become the spinal cord of modern enterprise. The emergence of new appropriate information tools satisfies the fast reactive business requirements (Zarli and Richaud, 1999). However, the so-called virtual enterprises (VE), which binds a fragmented and geographically spread set of collaborating partners together, increases the need for and elaboration of new powerful frameworks to support business models.

Monitoring changing user needs and requirements, along with shortening developments in the life cycle period, is a major concern for companies. Cooper and Wootton (1998) investigated the approaches to requirements capture from telecommunications and automotive and IT industries. It was found that the automotive company used market research agencies as an ongoing input into their development process, while computing companies were attempting to introduce a cross-disciplinary approach for both idea and requirements capture.

The following sections show the application of computer integrated environment (CIE) systems in the construction industry and how its features help rectify the issues in this industry by effective development and implementation. A few examples of the CIE systems developed for the construction industry will be explained in order to provide a clearer understanding of CIEs. The construction industry is very likely to be the most challenging sector for effective development and implementation of the CIE systems, because of its own characteristics and features. However, the concept of CIE as explained in this section does not mean that these CIE systems are limited to the construction industry only.

In this book, the construction industry has been selected as the implementation field because it is one of the most challenging business sectors. If effective CIE development and implementation can be achieved for the construction industry, it can be achieved for other industry sectors that are less challenging.

3.2 The Construction Industry and its Features

The construction industry is one of the major industrial employment sectors in Europe. In the past decade, construction companies have spent a great deal of effort and resources in improving their business processes. New forms of innovative project management, supported by recent IT developments, have evolved in response to ever-growing pressure from owners to complete projects on time and deliver high-quality buildings. Although many sectors such as automotive, manufacturing and service sectors have improved their competitiveness from IT, the construction industry has had some difficulties, resulting in the construction industry lagging behind the other sectors. The constraints the construction industry has experienced can be outlined as follows:

- **The nature of information and its flow:** Construction projects consist of many interrelated processes and sub-processes, often carried out by different professionals at different locations. Most of the tasks involved in construction processes are mainly about exchanging information between project stakeholders. All construction research studies have addressed the need to improve the poor cross-disciplinary communications, which will definitely lead to improvement in the efficiency and the effectiveness of the construction processes.
- **The construction industry and its role in the national economy:** Traditionally, construction companies have not fully perceived the importance of increasing the dynamism and complexity of its external environment. This could be attributed to the special and complicated nature of the industry and also to the lack of a long-term cooperative strategic thinking (Aouad and Wafai, 2002).
- **The fragmented supply chain of the construction industry:** One of the main features of the construction industry is the high fragmentation in its supply chain. However, despite the increasing trends towards multidisciplinary practical arrangements between construction firms, such as partnering, the construction industry still consists of hundreds of small and medium-size firms that offer undifferentiated products and services.

- **The culture of the construction industry:** The widespread culture of the construction industry is a demanding, confrontational one, which has underpinned the inefficiency and the ineffectiveness of its processes. The strong resistance to change could be partly attributed to the strong and rigid culture of the construction industry. However, recent research in the construction field has addressed the importance of cultural and people issues as the key and starting point for successful implementation of any innovative improvements in the construction processes.
- **Lack of long-term strategic management thinking:** The absence of sophisticated management techniques and methods is a dominant feature of common practices in the construction industry. Many research studies have highlighted the coherent lack of management expertise and the poor applications of strategic management in the construction industry.

Figure 3.1 depicts the traditional practices in construction projects. The construction industry is fragmented by nature. In addition to this fragmentation between the construction stakeholders, a high level of complexity of the work flow resulting from a high number of companies working on the same project increases the inefficiency of construction projects. For example, repeated processes or functions

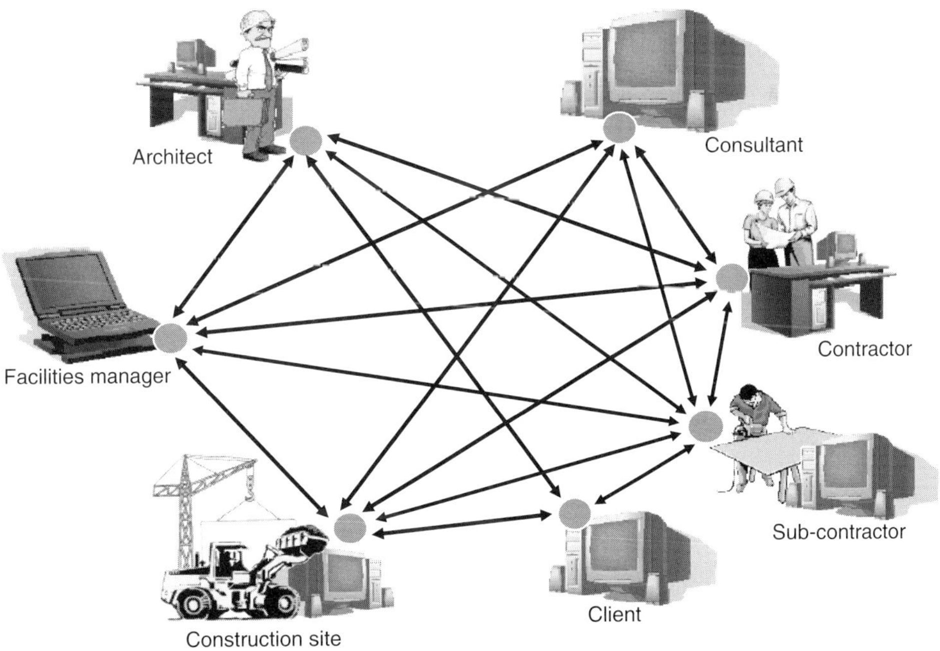

Figure 3.1 Current work environment in the construction industry.

and duplications due to the lack of communication and standardisation that causes waste and lead times in the project life cycle can lead to extra costs to the client for non-value-added activities.

- Design processes are separated from the construction processes, which increases the uncertainty and incompatibility between the design solution and the actual construction, i.e. poor buildability. High uncertainty caused by site conditions, as well as modifications in the projects that take place after the actual construction on site, has already led to discontinuity between the design and the project execution (Coudret *et al.*, 2001).
- Clients and some stakeholders such as local authorities and residents have an incorrect perception, or lack of understanding of the 2D architectural and engineering drawing. The design team cannot fully understand the client's needs because of a lack of communication, a shared platform and a virtual reality (VR) tool that is understandable by both the client and the architect. Eventually, these constraints will bring about client dissatisfaction.

3.2.1 Benefits of CIE to the construction industry

In the past, researchers have used IT tools to provide numerous decision support systems for the professionals of the industry. However, these systems have created islands of automation and are far from achieving an acceptable level of integration across disciplines and across the design and construction processes. It is recognised that greater benefits can be obtained and the above constraints can be considerably reduced if a complete integration based on VR tools is achieved. In this respect, major benefits of a desired integrated VR environment are considered as follows:

- Improving the coordination and communication among the client, design team members and construction professionals by using standard formats and intuitive VR tools that ease communication and information sharing
- Evaluating the design at the very early stages of the project life cycle since VR tools allow the design team to have quick and high-quality feedback on the project, in terms of architectural, technical, financial and environmental aspects
- Looking at 'what if' scenarios at the detailed design stage, to assess the design solution in lighting, acoustic and thermal aspects
- Closing the gap between the design team and the construction team and providing them with an integrated platform through

which they can collaborate for the best buildability and appropriate construction planning
- Using past project knowledge and information for new developments.

Furthermore, visualisation in conjunction with integrated model-driven construction systems can be expected to

- enable designers, developers and contractors to use the VR system and virtually test a proposed project before construction actually begins;
- offer 'walk-through' views of the project so that problems can be found and design improvements can be made earlier;
- provide a free flow of information between computer-aided design (CAD) systems and other applications work packages in order to minimise misinterpretation between project participants;
- facilitate the selection of alternative designs by allowing different plans to be tested in the same virtual world.

3.3 The Scope and Roles of CIE in Construction

The poor cross-disciplinary communications and the special nature of the industry and its supply chain have been regarded as the main bottleneck for any further improvements in the construction industry performance.

It was widely realised that the construction industry has much in common with manufacturing industries. Thus, construction research started to look at the construction industry from the perspectives of the successful models previously applied in manufacturing.

Computer integrated manufacturing (CIM), which is a customised term for CIEs for the manufacturing industry, has been presented as a key for integration in manufacturing. The CIM has led to the notion of computer integrated construction (CIC), which is a customised term of CIEs for the construction industry. The very basic premise of the CIC is the ability for different project participants to share project information either by accessing a central database or by exchanging information electronically.

Background research in construction shows that integration has been addressed in a variety of ways: communication between applications, integration through geometry, knowledge-based interfaces linking multiple applications, and multiple databases and integration through central databases, holding all the information related to a project according to a common infrastructure. The ongoing efforts in the area of integration and the use of IT in construction have led to several advances in the fields of Data Exchange Standards, which have been reflected in many research projects including Industry Foundation Classes (IFCs), International Alliance for Interoperability (IAI), Common Object Request Broker Architecture (CORBA) and Standard for the Exchange of Product Model Data (STEP).

Several prototype applications, addressing the integrated project data model and the implementation of the integrated project database, have emerged including ATLAS (Greening and Edwards, 1995), COMBINE (Augenbroe, 1995; Sun and Parand, 1998), COMMIT (Brown *et al.*, 1996), ICON (Aouad *et al.*, 1994), OSCON (Aouad, 1997), GALLICON (Aouad *et al.*, 2001), WISPER (Faraj *et al.*, 2000), SPACE (Alshawi *et al.*, 1996) and DIVERCITY (Arayici and Aouad, 2004), FIDE (Molina and Martinez, 2004), MOBIKO (Steinmann, 2004), PAMPER (Szigeti and Davis, 2003), BLIS (Laiserin, 2003) and nD Modelling (Aouad *et al.*, 2005).

3.3.1 Building information modelling (BIM)

Building information modelling (BIM) has become an internationally recognised concept to achieve by the development and implementation of CIEs in the construction industry. BIM can be described as the use of the information communication technologies (ICT) to streamline all the processes that require a building infrastructure and its surroundings to provide a safer and more productive environment for its occupants; to assert the least possible environmental impact from its existence and be more operationally efficient for its owners throughout the life cycle of the building infrastructure.

BIM in most simple terms is the utilisation of a database infrastructure to encapsulate built facilities with specific viewpoints of the stakeholders. It is a methodology to integrate digital descriptions of all the building objects and their relationships to others in a precise manner, so that stakeholders can query, simulate, and estimate activities and their effects on the building process as a life cycle entity. Therefore, BIM can provide the required value judgments that create more sustainable infrastructures, which satisfy their owners and occupants (Federal Facilities Council 2006).

BIM as a life cycle evaluation concept seeks to integrate processes throughout the entire life cycle of a construction project. The focus is to create and reuse consistent digital information by the stakeholders throughout the life cycle. Some advantages are (i) model-based decision-making, (ii) design and construction alternatives and

(iii) costs, energy and life cycle analysis. The automated building code checking (Cheng, 2005) or thermal load calculations (Kam *et al.*, 2003) or environmental impact assessment (e.g. carbon dioxide emission), etc. can be performed by using these models for design changes and alternative material types over the infrastructure life cycle.

Some attributes of BIM include robust geometry, comprehensive and extensible objects properties, semantic richness, integrated information and the ability to support the infrastructure life cycle (Schevers *et al.*, 2007).

BIM incorporates a methodology based on the notion of collaboration between stakeholders using CIE to exchange valuable information throughout the life cycle. Such a collaboration can be seen as the answer to the fragmentation that exists within the building industry, which has caused various inefficiencies. The above scenarios show the expected stakeholder collaboration and the aim of using BIM. Although BIM is not the salvation of the construction industry, a great deal of effort has gone into addressing those issues which have remained unattended for too long (Jordani, 2008).

The implementation of BIM systems requires dramatic changes in current business practices, which will lead to the development of new and sustainable business process models. BIMs in general and IFCs in particular are aimed at achieving interoperability among software tools that are used in the entire life cycle of a construction project. It is envisaged that all tools will be able to work on a central pool of project data. Although, the majority of architecture, engineering and construction (AEC) software developers have IFC APIs that are capable of importing and exporting IFC/STEP files, it is still not possible to make full use of the IFC model and abandon the file-based exchange scenario. This is attributed to the fact that an IFC model of a certain project is exchanged as a whole unit.

In the meantime, the internal structures of different software applications do not support the whole range of information that is covered by the IFC specifications. This makes it nearly impossible to maintain a lossless data exchange across applications (Nour, 2007).

3.3.2 Product models

At first, product models were used in data exchange between software applications bidirectionally between two applications. Recently in the pilot projects, the IFC standard has also been used as a software-independent, neutral data exchange format. Technically speaking, the IFCs are implemented in software applications to create a common data exchange platform. From a pragmatic viewpoint, the end user of the applications sees IFC as a new feature to save and retrieve data in a form that is also understandable to other software applications.

Product models have not only been used on a file basis but also on a model-server basis, where the participants' applications manage just their own share of the model data. The questions of version handling and design change management have so far been the most critical area in the first product model-server pilot projects. Technically speaking, a product model is not usually a single database: it is merely a collection of loosely linked databases, where the linking has been done with clear rules, also suggested to be one of the future development areas by Fischer and Kam (2002). The main benefits of the product modelling are as follows (Fischer and Kam, 2002):

- Object-oriented software and IFC shortened the design iteration time, expedited design on time and helped in developing and keeping a reliable budget.
- Less redundant design data was achieved (less need to re-enter geometric, thermal and material property data).
- The models supported an early design phase visualisation for project participants and improved collaboration among participants.

As the size and complexity of model information grows and as the global dispersion of design and production expertise grows, it is expected that model repositories will become more common for managing project workflows and data exchanges. The server technology is expected to provide a smoother transition, from a firm-based server capability to an off-site web-based capability.

Contemporarily, the software development visionaries extend their services into operations and maintenance support, leveraging BIM for the full life cycle of a facility, and moving towards offering competitive total cost of ownership (TCO) models (Jordani, 2008). This means that construction industry software providers will follow trends of other industries to offer Software as a Service (SaaS) model for collaborative project teams. The hosting of large data models for design and construction throughout the life cycle will be an important core competency for construction project delivery, which is expected from the ICT solution providers (Jordani, 2008).

3.4 Implementation of CIE in the Construction Industry

A consensus emerged on the subject of integration as a proposed solution, which has been highly praised by the research communities and the construction practitioners. The integration research studies could be divided into three main categories according to the breadth and depth of process and data integration: (i) electronic document management systems (ii) inter-operating autonomous systems and (iii) fully integrated concurrent engineering systems (Sun and Aouad, 2000).

Owing to an increasing complexity of product development and an intensifying market competition, VE appears to be a necessity in nearly all the industrial fields. Even large enterprises are no longer able to design and produce all the different parts of a product because of time constraints and the lack of some widely required specific expertise inside the enterprise.

All these factors indicate that there is a crucial demand in the construction industry for CIE solutions. CIE solutions enable the management of incompatible software applications that run on heterogeneous platforms, data exchange and interoperability mechanisms between applications, managing different types of information with different levels of performance and functionality, together with a powerful means of communication between distant applications. For an industry such as the construction industry, characterised especially by large numbers of SMEs and large groups of small businesses, this would lead to more agile production.

Over the last decades, there have been many indications of the increasing use of computer science and IT applications in construction firms. However, the evolution of computer science and advanced technologies in the construction industry has been much slower than in other manufacturing and servicing industries.

Most previous research has addressed the type and way of achieving the integration from a technological point of view. These efforts have led to the development of several technical solutions for the integration strategies. However, little attention has been paid to the issues in the implementation of these technologies (Aouad and Wafai, 2002). Furthermore, there is currently a need to develop methods and tools that facilitate the transformation of these sophisticated technologies into practical uses within the industry. In the following sections, some CIE examples are explained.

3.5 The CIE Case Study Project 1

Case Study 1 was a project that addressed the design fragmentation problems and provides a vehicle for storing architectural design information in an integrated object-oriented database shared across a range of computer applications developed within the project. One of the main aims of this case study is the assertion of construction industry requirements for an 'integrated solution' comprising common definitions of objects and functions, supported by software applications written for dedicated use in the construction sector. The CIE development achieved this integration by establishing standard object-oriented models that all applications can adopt and share.

3.5.1 The CIE system in Case Study 1

The CIE system is an interactive system for integrating CAD and construction-related applications to address the problems of design fragmentation and the gap that exists between construction and design processes. It provides a vehicle for storing architectural design information in an integrated construction object-oriented database that can be shared by a range of computer applications.

Hybrid information and an object-oriented approach were applied to establish the principles for integration. The integration was based on an applications-independent, object-oriented database and an architecture for providing access to that database through a variety of common and proprietary application software packages. It adopted a model-based approach to capture graphical and textual information about building components and directly stored them into an object-oriented database, rather than the complex and ill-structured CAD databases.

This approach avoids the expensive process of translating drawing data from the CAD database into an object-oriented database. Since the system complies with IFC, it facilitates data exchanges between construction-related applications and allows a common interpretation of the design objects among the applications in a consistent and interoperable manner. Regarding project data interchange, it provides sharing and inter-working between objects distributed among applications, by complying with CORBA standards. The following are three important aspects addressed by the CIE system in Case Study 1:

- An object-oriented database for CAD drawings
- The independence of the architectural design object from the display environment
- The use of the IFC and CORBA standards to promote a collaborative working environment.

The system design was built using the Unified Modelling Language (UML). The use of Rational Rose CASE tools was helpful because the tool can automatically generate both C++ code and CORBA Interface Definition Language (IDL).

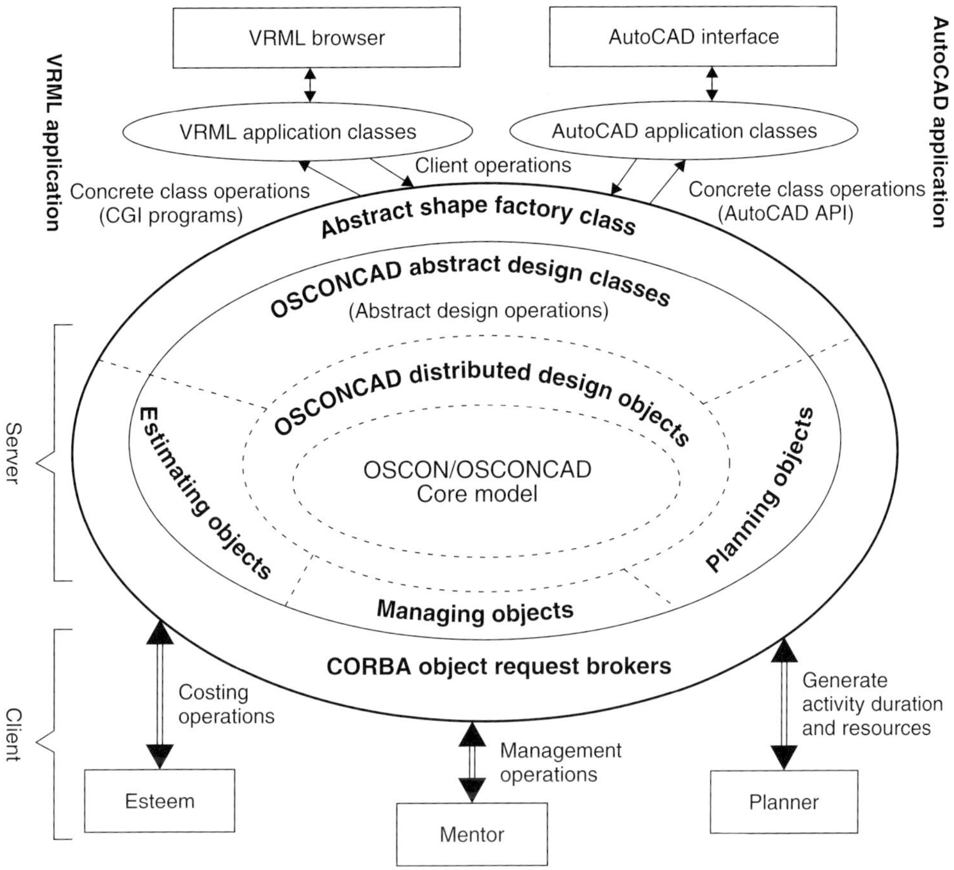

Figure 3.2 The system architecture of CIE in Case Study 1.

The system architecture

The system architecture adopts the client/server approach as shown in Figure 3.2. The server is composed of the following two kinds of objects:

- Objects that reflect the database schema of the architectural design model where instances of the design component are stored
- A set of abstract design classes, which provide abstractions that the design model classes can use to render themselves in a given display environment.

The AutoCAD and VRML are the front-end applications. They use the abstract design classes to derive concrete classes to display instances of the design in the AutoCAD and VRML environments. The cost estimator (Esteem), the project manager (Mentor) and the project planner (Planner) applications are the client applications to access and manipulate instances through CORBA object request brokers. The system was designed to be generic and abstract enough to allow separation between the design model and the specific implementations of commercial CAD tools. For this, it uses the Abstract Factory Design Pattern, which provides an interface for creating families of related objects without specifying their concrete classes.

The AutoCAD application

The application is an AutoCAD Application Development System (ADS) prototype. It allows the user to create and manipulate architectural components of a building. The created components are stored as instances of classes of the application design model in the object database through which they are made available to other AEC computer-aided applications in computer-interpretable and standard form.

The application provides a simple environment to design building components. The design primitives are walls, windows, doors, columns, beams, foundations, etc. As design objects are created through the AutoCAD interface, corresponding objects are created in the object database. The information that is used to draw these objects on the AutoCAD screen via an instance of the *Drawer* class comes directly from the instances in the database. This ensures that the screen reflects the design database. This also means that changes in a design can be immediately propagated to other areas of the project database. For instance, if an online project-planning package accesses the database, then it would be possible to change a part of the design and immediately see the impact that these changes would have on the project schedule. The prototype provides the following functionalities:

- Creation and manipulation of foundations, walls, windows, doors, floors, stairs and roofs
- Creation of space diagrams and expansion into building elements
- Creation of specifications and specification sets for application to a design
- Texture mapping of building elements based on component specifications and finishes.

A command-driven model of interaction was chosen for the application. This is the standard way of implementing an AutoCAD application, with extra sets of menus being added to the standard

AutoCAD menu hierarchy to invoke the new commands. These menus provide commands to create and manipulate building elements such as walls, windows, foundations, etc. Levels have been defined for the design components so that they can be drawn at the locations specified by the user. The user has the option to draw in 2D or 3D. Editing facilities have also been provided. For instance, the user can move junctions, delete and copy objects, and any changes made to the design are reflected in the object database.

The VRML application

This application is a web-based application using Virtual Reality Modelling Language (VRML). It was used as a means of remotely interrogating information stored within the integrated database and allowing the user to visualise and manipulate the design components in a 3D environment. Based on the design information stored in the object database, the VRML_Drawer generates a programme for the VRML environment. The generated programme includes nodes such as a WWW anchor (which allows links to web pages), the material, rotation, and the cube and is used by the VRML browser to create a 3D environment that can have links to web pages containing information about the design components.

In the CIE system, VRML is used not only as a visualisation tool but as a user interface as well. For instance, the user could interact with a 3D column in VRML rather than a column in a traditional database environment. This will allow the construction practitioners to have better access to information, which will motivate them to use integrated databases.

The user can navigate inside the building clicking on design objects and retrieving information about their properties, which include geometric, cost and time data. This direct interaction with a VR environment has many advantages over the use of a CAD package. The ultimate benefit is the ability to create a walk-through, which can facilitate collaboration among clients, designers, contractors and suppliers. Such collaboration is the main objective of the database and VR is a powerful medium for communication and convergence.

Applications linked to the AutoCAD application

As shown in Figure 3.2, the major components are the core integration model (Aouad *et al.*, 1997), the models for the architectural design, cost estimation and construction planning domains. From these conceptual models, a distributed object database schema is established as a means of integrating the information used by a number of participants within a construction project. The design instances created in the database provide different construction experts with the information they require and serve as the main medium for the integration in Case Study 1.

To demonstrate the feasibility and practicability of this integration approach, three construction applications have been developed to access and share the building design instances generated by the AutoCAD application of the CIE system and stored in the object database (Aouad *et al.*, 1997) – Esteem for cost estimation, Mentor for project management, and Planner for construction planning. As a result of the use of the CORBA standard, other construction applications could be plugged in to access the distributed objects.

- **Esteem:** Esteem is an object-oriented application developed after producing an object model that incorporates object types such as resources, work items, work specifications, etc. Within Esteem, the user can change the rate, productivity, etc. of a resource; the information is then changed in the database. Once a design is created, Esteem will read that design and create instances of work items in the database and populate the attributes of quantities and rates. Quantities are derived from the design components, whereas rates are established from the library of a company's work items and resources. Esteem can then establish the building project's total cost. Any changes to the design model are notified to objects within the Esteem costing model in order to ensure consistency of information. Esteem contains a list of work items and quantity and rates generated automatically by the object-oriented database. The user can view the resources associated with work items and change some of the costing parameters such as price and productivity.
- **Planner:** Once the design module creates a design, a series of construction tasks associated with that design are instantiated in the object-oriented database as there is an object model that supports the planning application. These tasks contain information about building duration. They also have attributes about the starting date, the finishing date and dependency, which are populated once the plan is processed by the super-project planning application. File import/export is used to update the information in the database. Any changes to the design model are notified to the planning system in order to ensure the consistency of the information. The user can change the resource requirements and dependency.

- **Mentor:** Mentor is an object-oriented process management system. It contains objects such as process, processor, rights, etc. The concept 'process' is itself treated as an object. The users can instantiate processes in the object database. He or she can freeze, delete or update, etc. that process. The object-oriented model supporting Mentor is fully integrated into the other models and applications. For instance, if the process design is instantiated in the database with a freeze operation, this will disallow the AutoCAD design application to operate. For a comprehensive description of the aforementioned applications refer to Aouad *et al.* (1997).

3.6 The CIE Case Study Project 2

Case Study 2 was a research project funded by the Department of the Environment, Transport and the Regions (DETR) of the United Kingdom under the Information Technology Construction Best Practice Programme. The aim was to demonstrate the benefits of applying integration technologies to the water and housing sectors of the construction industry. Two integrated systems have been developed in this project (Marir *et al.*, 1998):

- The first system is a distributed environment that considerably improves the conceptual design of wastewater treatment plants and minimises development costs by integrating the process of design, cost estimation and project planning.
- The second system is a back office application that creates a website for customer relationship management in housing projects by integrating the process of design, cost estimation and project planning.

Both systems have been tested in collaboration with the aforementioned companies and received a very positive response. The aims and objectives are outlined below and Figure 3.3 illustrates the integrated project model:

- Easily accessible data being shared among the client, contractor, quantity surveyor and the design professionals
- Direct input of supply data by key suppliers to provide certainty of specification and price throughout the supply chain
- Resolution of build problems prior to site commencement by effective information sharing at the briefing stage
- Improved site efficiency through better resource coordination
- Improved reuse of data typically lost during the construction process
- Enhanced trust through transparent information flows
- Streamlined administration of the supply chain.

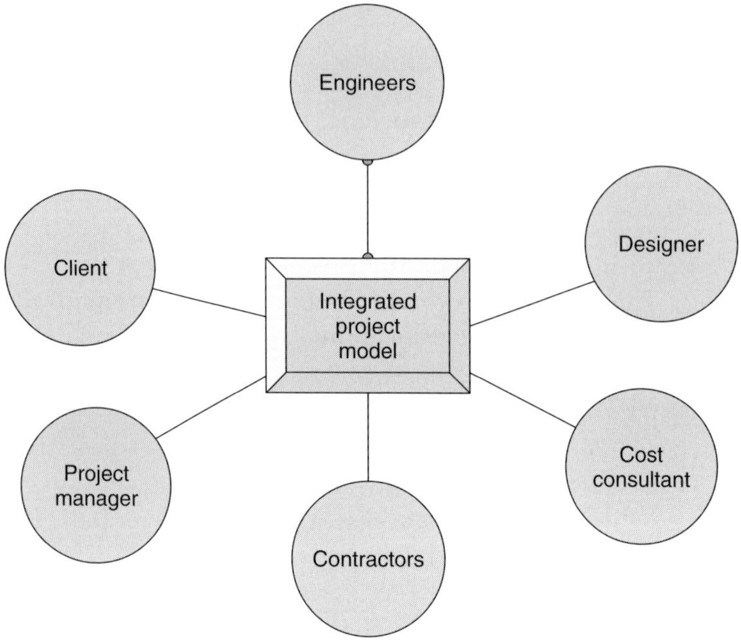

Figure 3.3 Integrated project model of the CIE system in Case Study 2.

3.6.1 The CIE system in Case Study 2

The prototype for wastewater treatment projects and a prototype for housing projects were developed during the project. The purpose of these systems was to assess the implications of the integrated project database approach to real projects, to verify the validity of the process and data models developed during the requirement analysis phase and to investigate and address technical issues related to the development of systems.

The wastewater treatment prototype was an outline design tool developed to support the early design stages of wastewater treatment projects. Its aim was to support partnering arrangements where distributed teams work together to develop the scheme definition of a feasibility study. It consists of a CAD, a cost estimator, project planner and a client management application. These applications operate in an integrated manner; changes in one of them are reflected by changes in the others.

This means that, for example, by creating a design in the CAD application, the cost estimating and project planning applications automatically generate cost and planning information related to that design. The purpose of the CAD, cost estimating, and project planning applications is to improve the scheme definition development process.

The purpose of the client management application is to improve the communication between the development team and the client in order to assist the client to play a more active role in this process. This is done by enabling the client to access layout, cost, and planning information of the developed design over the Internet. This information is generated automatically on the basis of the design, cost and planning information that already exists in the system and is presented in a VR environment implemented using the VRML.

The housing prototype is a tool that automatically constructs a website from input given in the CAD and project planning applications, which again operate in an integrated manner. The purpose of this site is to improve the relationship with customers. Through it, the customer can view the layout of a housing site in a VRML environment and reserve a house over the Internet. Customers who have already bought a house can customise it through the site by ordering various extras.

The system automatically determines when an order can be accepted, depending on the progress of the house, the available customisation options, and the availability of the items related to the order. Information about the progress of the house is found in the integrated database while the designer in the CAD application specifies the available customisation options. Accessing suppliers' databases and checking the available stock determine the availability of the items related to the order or the delivery dates of new stock, which must agree with the house construction time plan for the order to be accepted.

Integration in the CIE system in Case Study 2

Integration of the CAD, cost estimating, project planning and client management applications in Case Study 2 are based on a central database that is used by the applications to share data. Each one of the applications generates part of the data related to a specific entity, such as a construction object. These data are stored in the central database in a common format that is rich enough to accommodate the needs of all the applications. This format is based on the data model that was developed during the requirement analysis phase of the project. Changes in the data by one of the applications trigger changes to the corresponding data in the others. The following diagram in Figure 3.4 shows the flow of this interaction.

The CAD application is the starting point where new construction objects are created or existing ones are modified. Construction objects are created and drawn in the CAD application by calling built-in commands, and not by using basic shapes such as lines and circles. For

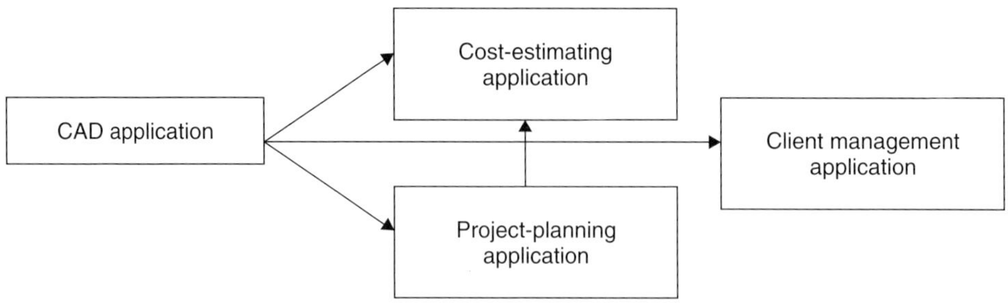

Figure 3.4 Integration flow in the CIE system in Case Study 2.

example, to draw a nitrifying filter in the wastewater treatment prototype, the designer calls the command 'Draw Nitrifying Filter'. He then specifies a series of attributes, such as the filter's position, geometry and composition, and the system accordingly draws a representation of the filter.

In Case Study 2, this representation is the footprint of the construction object. This is because the footprint design of a construction project is adequate for the purposes of both prototypes. However, it should be noted that although a 3D representation of each object is not shown in the CAD application, information about its 3D geometry is captured and stored into the system.

Each object in the CAD application is an information-rich object that has specific attributes and behaviour. It knows its role in the design and its relationship with other objects. For example, each object knows its possible connection points with a pipe, if it can be connected to one. To connect two objects with a pipe the designer has to select one of the possible connection points presented by the system and not explicitly specify the coordinates of the connection points. When he moves one of the objects, the system accordingly readjusts the pipe so that the two objects remain connected.

Each time a new object is created in the CAD application or an existing one is modified, the cost-estimating and project-planning applications read its properties and perform various calculations. By having access to detailed information about each object and related external factors, such as the geology and topology of the project site, the cost-estimating application calculates various costs by applying quantity surveying rules. However, it should be noted that the accuracy of the calculations varies, depending on the type of costs being calculated. This is because for some type of costs the calculations depend on factors that are difficult to specify or require information that is unavailable at the stage of the construction project where the system is used.

Case Study 2 did not aim to automatically generate complete and almost accurate cost estimation. It rather aims to assist the user in calculating some of the costs, to assist the user to perform a what-if analysis where relative costs are more important than absolute ones and mainly to improve the information flow between the project partners.

The information flow is improved either by automatically calculating costs after making changes in one of the applications, or by informing about the changes that have been made and providing all the necessary information that is needed to calculate the corresponding costs manually. The user has the overall control in the cost estimation and any changes that he makes are fed back in the central database.

In a similar way, the project planning application does not generate a complete project plan for the construction project. Generating a project plan automatically is even more difficult than generating cost estimation, since a plan can depend on many unpredictable factors, such as the availability of some staff.

What the CIE system in Case Study 2 did generates a list of all the tasks associated with a project and their durations. The project manager has the overall control to accept or change the generated tasks and to create other critical paths. Once again, the system assists in the exchange of information by performing any changes that can be done automatically or by informing about those that must be done manually. Any changes that are made by the project manager are fed back into the central database and reflected to the other applications.

Most of the cost and task duration calculations in the CIE system are based on the unit costs of the materials, including the construction objects and the unit costs and durations of any tasks associated with them. In Case Study 2, most of these costs and durations are taken from the Contractor's Price Database. A connection with a different database is also possible.

Architecture of the system in Case Study 2

An integrated system would be of no real value if new construction objects could not be easily added to the system and existing ones could not be easily modified. Object-oriented software development gives the opportunity to develop systems as collections of many independent components called *objects* that are tied together and controlled by 'driving' applications. By adding new objects to the system and modifying existing ones, we provide a new functionality without the need to modify the driving applications.

Case Study 2 follows the object-oriented paradigm and, as such, an object-oriented database is more suitable as the integrated database. ObjectStoreTM version 5.0 was chosen as the object-oriented database. The CIE system in Case Study 2 consists of a library of objects that have been developed according to the data model, a number of external applications that handle all the user interactions and a number of interface applications that connect the external applications to the central database where all the information is stored. Each object in the library knows how to draw itself in the CAD application, how to calculate any costs and tasks associated with it and its relationship with the other objects.

The architecture of the CIE system in Case Study 2 is shown in Figure 3.5. This architecture separates the user interface functionality from the core functionality, thus enabling changing the external applications in a plug-in fashion. Once an appropriate interface has been built, any application can be used to handle the interaction with the user. It should be noted that all the data are stored as objects in the central database and not in the application's native format.

For the CIE system, AutoCAD was chosen as the front-end of the CAD application of the system, Microsoft Excel was chosen as the front-end of the cost estimating application, Microsoft Project was chosen as the front-end of the project-planning application, and a web browser with a VRML client as the front-end of the client management application. The AutoCAD ADS library was used to control AutoCAD, the Microsoft Component Object Model (COM) technology was used

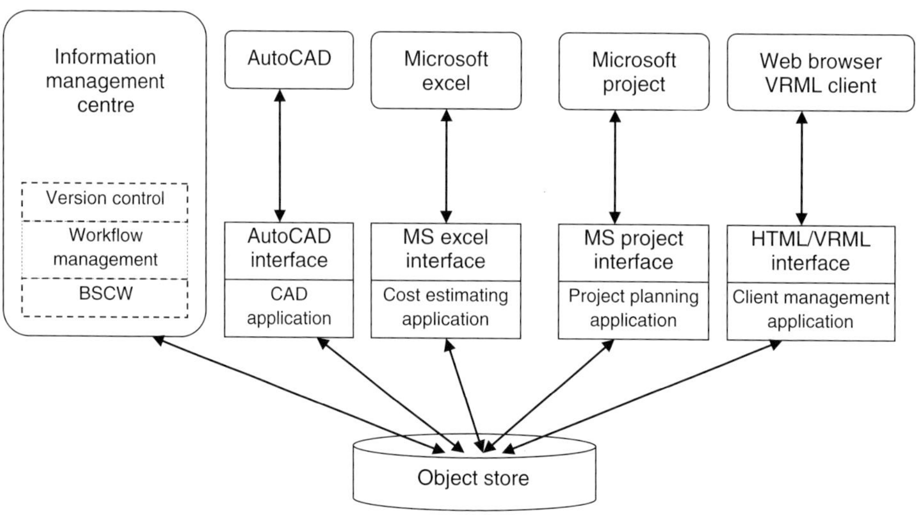

Figure 3.5 The CIE system architecture.

to control Excel and MS Project and the Common Gateway Interface (CGI) to control the web browser and the VRML plug in. All the interface applications have been built using the Microsoft Visual C++ version 5.0 programming environment.

3.7 The CIE Case Study 3

Case Study 3 was on distributed virtual workspace for communication in construction projects and was an EU-funded project. The project used IFC standards in order to develop a toolkit for shared virtual briefing and design in the construction industry. This toolkit allows construction companies to conduct client briefings, design reviews, simulate 'what if' scenarios, test constructability of buildings and communicate and coordinate design activities between teams. Both synchronous and asynchronous interaction are emphasised in this framework.

The CIE system in Case Study 3 allowed users to produce designs and simulate them in a virtual environment. The designs were IFC based and viewed by all stakeholders within the project team. The project had the following objectives:

- Creation of a client-briefing workspace, which can facilitate interaction and communication of design ideas between client and the architect
- Creation of an interactive design review workspace as in Figure 3.6, which can facilitate multidisciplinary design reviews involving different stakeholders of a construction project, i.e. planners, architect, designers, civil engineers, electrical engineers, contractors, facility managers, security personnel, etc.
- Creation of a virtual construction workspace, which can assess the buildability (the sequence of construction activities, scheduling, material handling, etc.) of a building
- Specification and development of a software framework for integrating the above three workspaces and sharing them over networks to support collaboration between geographically distributed project team members.

3.7.1 The CIE system in Case Study 3

Organisational changes of collaboration forms take place globally in the United Kingdom and in Europe with a focus on dismantling discipline borders between the building process parties. The CIE system

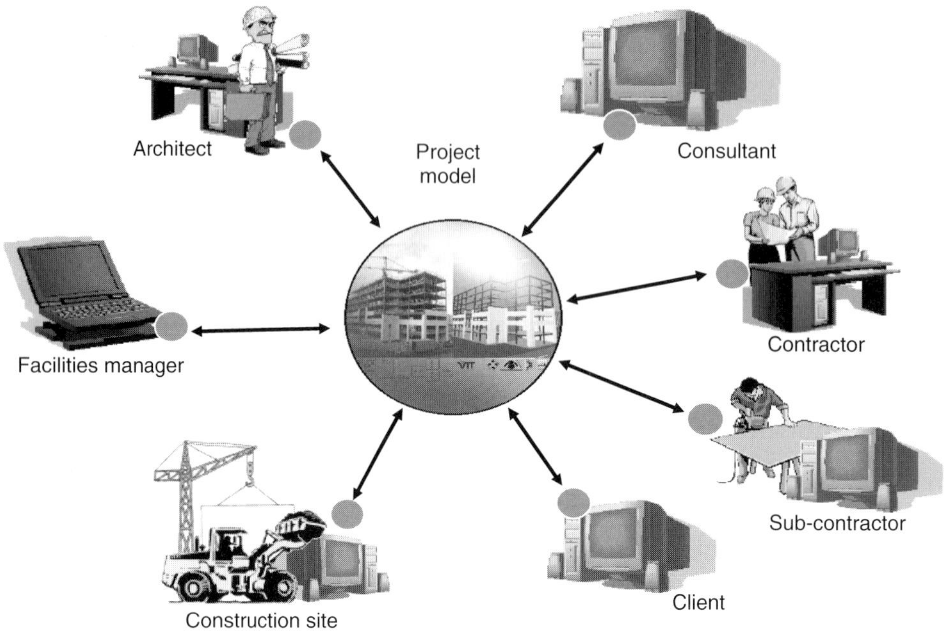

Figure 3.6 The ideal work environment via the CIE system in Case Study 3.

is a long desired tool to support collaboration, both across and along the building process time axis. Cultural barriers within the building process will hopefully be less important with efficient communication tools with a high degree of realism and a distributed presence through multimedia/VR and 3D interfaces. Added to the efficiency gains, a common understanding of building process and product quality issues may also be developed (Christiansson *et al.*, 2002).

To achieve its aims, in Case Study 3, six VR-based application prototypes were developed, namely, Client Briefing, Lighting Simulation, Acoustic Simulation, Thermal Simulation, Site Planning and Analysis and 4D Construction Planning Simulation. Added to these six applications, a communication layer for stakeholder collaboration and information storage, and the integration framework for integration of the above seven applications including the communication layer were also developed in the project.

The CIE system was comprised of the following applications:

1. **Client-Briefing Application:** This represents the interface between the client and design team. It provides tools for capturing strategic and spatial requirements and allows experimentation with alternate spatial layouts in a visual format (Aspin and Fernando, 2002; Patel *et al.*, 2002). It is intended to produce a single initial design, agreed

on by all parties, and as this design iteratively and progressively turns into a formal detailed design, feedback is obtained in order to drive the design process forward.

2. **Design Review Applications:** Although state-of-the-art software tools exist for the detailed design stage, throughout the user requirements capture in the project it has been observed that these existing tools suffer from two important limitations (Shelbourn *et al.*, 1999):
   - Discontinuities between the different software tools. This makes reuse of the results of one technical domain as an input for another technical domain practically impossible.
   - Lack of 3D real-time inspection features. Consequently, members of the project team spend too much time trying to (i) understand the project information and (ii) describe this information to another one.

   In order to greatly reduce the above limitations, an interactive design review workspace that allows the project teams to visualise and interact with the project on a multidisciplinary level was created. The main features of the design review workspace are as follows:
   - Supporting continuous design between different phases and within the detailed design phase using IFCs, which means that the calculation results yielded for one technical domain can be reused as an input for another technical domain.
   - Model-driven approach that allows project teams to share the same view about the project through a visual and shared conceptual model. As a result, substantial improvements can then be made on the communication level between the project teams.

   The activity diagram in Figure 3.7 shows the integration and collaboration between applications within the CIE system in Case Study 3.

   Design review applications within the CIE system were as follows:
   - **Lighting Simulation:** This provides a realistic simulation of lighting. The users can change and move objects or lights in the building and see lighting simulation in an interactive manner (Thery *et al.*, 2001). Furthermore, one novel characteristic of this application is the mathematical algorithms, which are used for simulation. Lighting simulations generally take a long time (several hours). In most cases, algorithms allow for real-time simulation and only take a few seconds. This facilitates interactive and collaborative simulation (Gobbetti *et al.*, 2002).
   - **Acoustic Simulation:** This application automatically reads CAD drawings (in an IFC format) and allows changes to be made

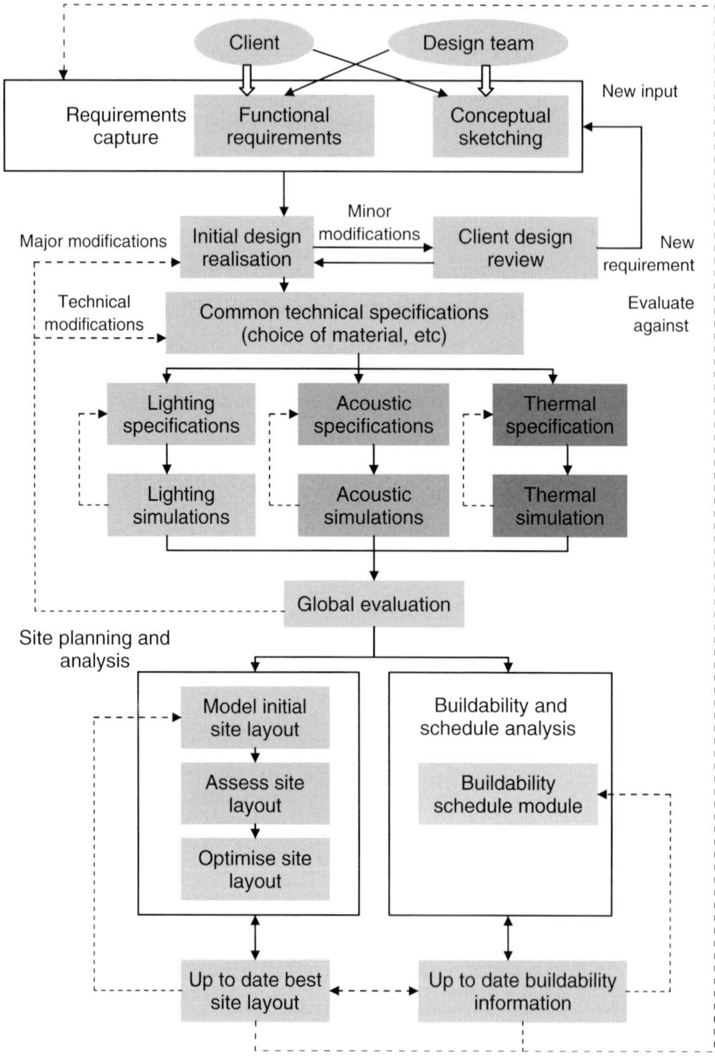

Figure 3.7 The CIE system activity diagram in Case Study 3.

to building materials online in an interactive manner. The user can listen to the acoustic qualities of the building and change the design to achieve desired noise levels (Coudret *et al.*, 2001; Marache *et al.*, 2002).

- **Thermal simulation:** This offers the users the ability to automatically read the CAD model yielded by a CAD tool supporting IFC export, to interactively change materials of the building components (walls, floors, etc.) and to simulate variation of temperatures in different rooms and calculate exploitation costs (Coudret *et al.*, 2001; Marache *et al.*, 2002).

3. **Construction Workspace Applications:** The construction workspace is aimed at providing functionality to allow for rehearsing, evaluation and optimisation of the construction planning stage. It can be thought of as testing the constructability of a building by assessing both temporal and spatial aspects resulting from a planned schedule so as to identify and resolve potential conflicts that would otherwise impose high costs if treated at later stages (Fernando *et al.*, 2001).
   - **Site Planning and Analysis Application:** This application aims to design a modelling and simulation platform for supporting the construction site analysis stage and allows the safety evaluation and optimisation of the construction site layouts. In particular, it addresses the space planning aspects by assisting with the representation and management of spatial requirements in the construction site (Tawfik and Fernando, 2001).
   - **4D Construction Planning Simulation:** This is a 4D VR simulation application that shows step by step how the progress of a construction project will look in practice. The application links a standard IFC-based 3D building model with an associated construction schedule, which can be prepared with an off-the-shelf project management software package. This helps in the visualisation of the status of the building and its components at the selected point of time (Kähkönen, 2001).

   Information exchange among the applications is possible via IFC standards.
4. **Communication Layer:** The communication layer is at the heart of the distributed features of the CIE system. It provides support to allow virtual collaborative spaces and geographically distant sites to work together.

The communication layer of the CIE system in Case Study 3 employed XML as the distribution layer for the exchange of information (Figure 3.8). The implementation of the communication layer is based on Simple Object Access Protocol (SOAP) Internet protocol. One of the main advantages of SOAP protocol is that it deals with proxies and firewalls, which are often very strict in the industry domain (Da Dalto and Gobbetti, 2001). The communication layer provides the following:

- Communication among heterogeneous systems, architectures and languages
- Robust and secure message transfer and information storage

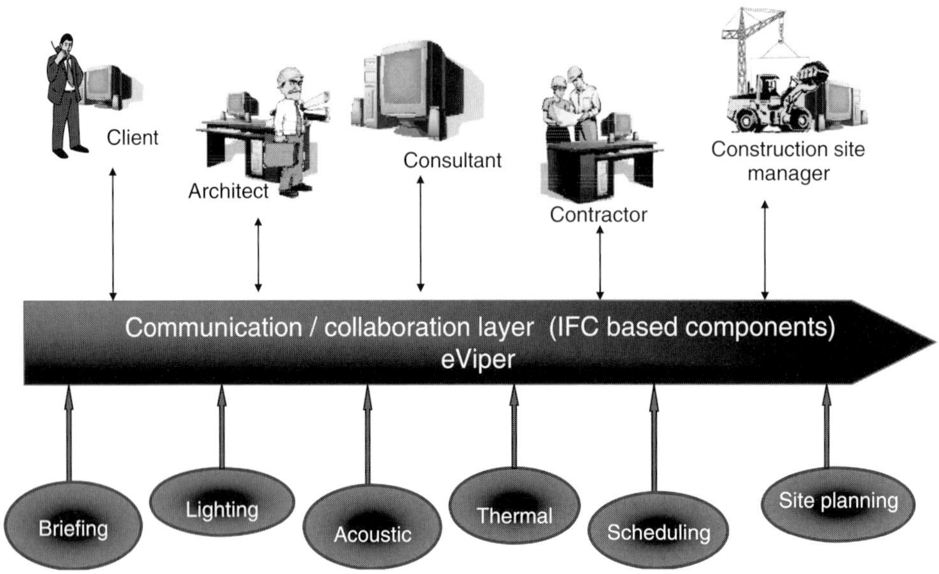

Figure 3.8 The CIE system work environment in case study 3.

- Time performances to allow real-time collaboration (only for specific messages – 3D scene motions and updates)
- Multi-user management including identification and access control.

The CIE system architecture in Case Study 3

In order to provide a single system architecture, which would support all the applications, the CIE system incorporated three key features in its architecture: (i) distributed database, (ii) configurability, and (iii) extensibility. The CIE system in Case Study 3 was a client/server system, implementing a traditional three-tier structure, as illustrated in Figure 3.9.

The server-side functions included the maintenance of a project database and configuration management (version control) of the project data. A single client application was developed acting as a framework for integrating software components, each of which implements a discrete software tool. The client application was described as a *Workspace Manager* in that while it implemented any user functionality, it provided a framework for the integration of the functionalities selected by the user, presenting them as a single integrated workspace.

The Workspace Manager allowed third-party developers to implement components that would extend the CIE system functionality. These third-party components might either be additional tools or be

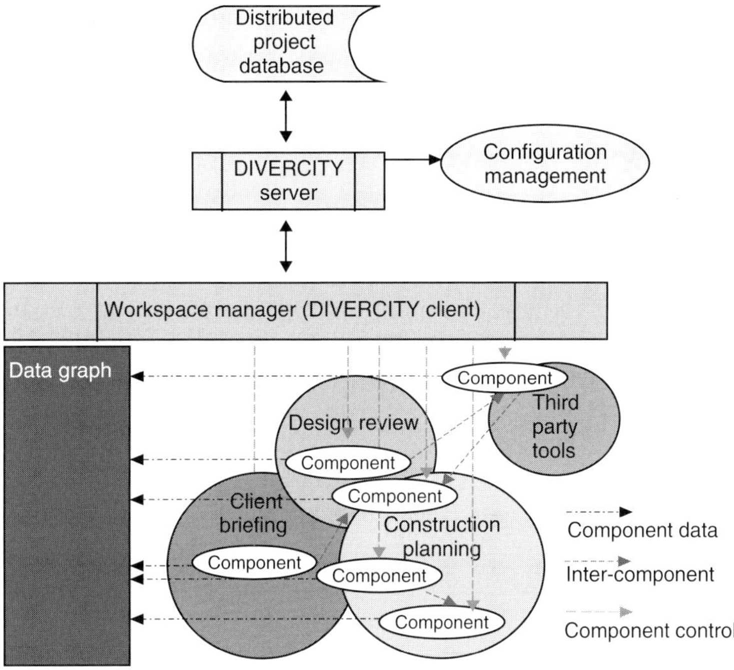

Figure 3.9 The CIE system architecture in Case Study 3.

interfaces that allow existing software applications, such as CAD, to interact with the CIE system. Third-party applications would then use an interface component to interact with a running client application, which in turn would interface with the server-side database.

A single client application was developed to act as a framework for integrating software components, each of which implements a discrete software tool. The client application was described as a Workspace Manager, in that while it implemented any user functionality, it provided a framework for the integration of the functionalities selected by the user, presenting them as a single integrated workspace.

Case Study 3 also supported VR display technologies, such as CAVE, Reality Room and ImersiDesk. Applications capable of utilising these devices were implemented as either stand-alone applications, which communicate with the client application in the same manner as third-party software, or as additional components within the client applications' toolset.

In order to address data issues at run-time, particularly that of memory usage, it employed a Data Graph, which was a single, centralised, shared, run-time data repository, residing in the Workspace Manager. Components will also be able to register interest in elements contained within the Data Graph, thus receiving notification of changes in the data; it is anticipated that this would form the main communication mechanism between components. Each component of the system was responsible for creating its own communication network with other components. To achieve this flexibility each component defined three communication mechanisms:

- **Component Control:** This was a one-way communication from the Workspace Manager to the component to load and unload the component from the client application.
- **Data Access:** Each component formed a link to the Data Graph to access application data. Application data might then be changed directly via this link. Components registering interest, in particular data elements, might also be notified of changes that may have been made to that data. This forms a shared data communication mechanism that will be the preferred inter-component communication technique.
- **Inter-component:** Components must be capable of forming networks with other components. Once formed, these networks will allow one component to access the exposed functionality of another.

3.8 The CIE Case Study 4

Case Study 4 aimed to enable and equip the design and construction industry with a tool that allows users to create, share, contemplate and apply knowledge from multiple perspectives of user requirements. This project differs from 4D modelling tools as its objective is to develop infrastructure, methodologies and technologies that will facilitate the integration of time, cost, buildability, accessibility, sustainability, maintainability, acoustics, lighting and thermal requirements, which is called *BIM* contemporarily. The tool will allow construction professionals to perform a true 'what-if' analysis at a very early stage of a project, based on the manipulation and impact of changes to the previously mentioned parameters, so that informed decisions can be made.

The terms *nD modelling* and *nD CAD* are becoming escalating idioms associated with ICT-based building design. An nD model is an extension of the building information model that incorporates multiple aspects of design information required at each stage of the life cycle of a building facility. The issues surrounding nD-enabled construction are not just technological but also cultural, human and process related (Aouad *et al.*, 2005). The concept of nD modelling builds upon the concept of 4D modelling by further integrating an nth number of design dimensions into a holistic model, thus enabling users to portray and visually project the building design over its complete life cycle (Lee *et al.*, 2003).

Both graphical and non-graphical documents, such as drawings and specifications, schedules and other data respectively, are included. Changes to each item are made in only one place and so each project participant sees the same information in the repository. By handling project documentation in this way, communication problems that slow down projects and increase costs can be greatly reduced. Leading CAD vendors such as AutoDesk, Bentley and Graphisoft have started to heavily promote BIM. However, as these solutions are based on different, non-compatible standards, an open and neutral data format is required to ensure data compatibility across the different applications.

IFCs, developed by the IAI, can provide such capabilities. IFCs provide a set of rules and protocols that determine how the data representing the building in the model are defined, and the agreed specification of classes of components enables the development of a common language for construction.

IFC-based objects allow project models to be shared whilst allowing each profession to define its own view of the objects contained in that model. This leads to improved efficiency in cost estimation, building services design, construction and facility management: IFCs enable interoperability between the various AEC/FM software applications, allowing software developers to use IFCs to create applications that use universal objects based on the IFC specification. Currently, most of the major CAD systems provide support for IFC export and this will ensure that the data is consistent and coordinated. Furthermore, this shared data can continue to evolve after the design phase and throughout the construction and occupation of the building.

3.8.1 The CIE system in Case Study 4

In Case Study 4, the CIE system development project developed a holistic nD modelling tool using IFCs to help improve the decision-making process and construction performance by enabling a true 'what-if' analysis to be performed to demonstrate the real cost in terms of the variables of the design issues. Therefore, the trade-offs between the parameters can be clearly envisaged, which will help to do the following:

- Predict and plan the construction process.
- Determine cost options.

- Maximise sustainability.
- Investigate energy requirements.
- Examine people's accessibility.
- Determine maintenance needs.
- Incorporate crime-deterrent features.
- Examine the building's acoustics.

The aim in Case Study 4 is to enable an informed design decision based on a variety of design perspectives. Traditionally, a whole host of construction specialists are involved in instigating the design of modern buildings. With so much information and from so many experts, it becomes very difficult for the client to visualise the design, any changes applied and subsequent impacts on the time and cost of the construction project. Changing and adapting the design and planning schedules and cost estimates to aid the client's decision-making can be laborious, time consuming and costly.

Each of the design parameters that the stakeholders seek to consider will have a host of social, economic and legislative constraints that may be in conflict with one another. Furthermore, as each of these factors varies in the amount and type of impacts they can have, they will have a direct impact on the time and cost of the construction project. Therefore, the criteria for successful design will include a measure of the extent to which all these factors can be coordinated and mutually satisfied to meet the expectations of all the parties involved (Aoaud *et al.*, 2005).

Specialist design criteria input is usually undertaken in a sequential step-by-step manner, whereby after satisfying the legal requirements, the design undergoes a number of changes; it then proceeds to the next consultant, who in turn makes a number of design recommendation changes. Design changes are made in isolation from each other in an over-the-wall manner, whereby each discrete change has little or no regard to the next. Therefore, it is often difficult to balance the design among aesthetics, ecology and economy – a three-dimensional view of design that acknowledges its social, environmental and economic roles – in order to satisfy the needs of all the stakeholders.

These problems are mainly attributed to the vast amount of information and knowledge that is required to bring about good design and construction coordination and communication within a traditionally fragmented supply chain. The complexity of the problem increases with the fact that this information is produced by a number of construction professionals of different backgrounds. Therefore, without effective implementation of IT and processes to control and manage this information, the problem will only intensify as construction projects become more and more complex and as stakeholders increasingly enquire about the performance of buildings (sustainability, accessibility, acoustic, energy, maintainability, crime, etc.).

The key feature of the system is its ability to incorporate numerous design perspectives in one system and its subsequent capacity to systematically assess and compare the strengths and weaknesses of different design scenarios presented by the nD knowledge base. The system architecture is illustrated in Figure 3.10. The tool was built on the concept of BIM and was IFC based:

- **nD knowledge base:** This is a platform that provides information analysis services for the design knowledge related to the various

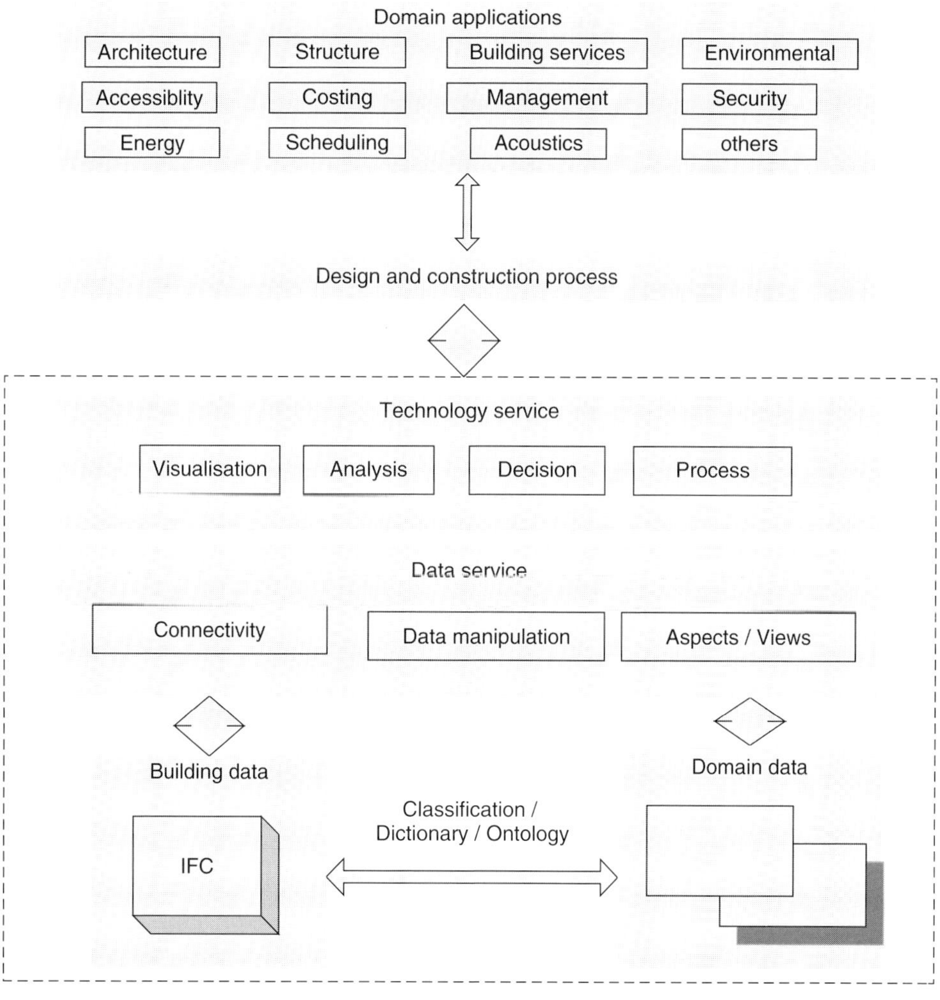

Figure 3.10 System architecture of the CIE system in Case Study 4.

design perspective constraints (i.e. accessibility requirements, crime deterrent measures, sustainability requirements, etc.). Information from various design handbooks and guidelines on the legislative specifications of building components will be used together with physical building data from building information model to perform individual analysis.

- **Decision support:** Multi-criteria decision analysis (MCDA) techniques have been adopted for the combined assessment of qualitative criteria (i.e. criteria from the Building Regulations and British Standard documents that cannot be directly measured against in their present form) and quantitative criteria (e.g. expressed in geometric dimensions, monetary units, etc.). Analytic Hierarchy Process (AHP) is used to assess both qualitative criteria (i.e. criteria that cannot be directly measured) and quantitative criteria (e.g. expressed in dimensions, monetary units, etc.).

The CIE system architecture

The system architecture was identified for different domain applications, which are directed by a business process and was also based on a service-oriented platform supported by various technologies such as visualisation, decision support and analysis:

- The system architecture was based on the service-oriented architecture (SOA). A common framework and interface was provided so that applications from different domains can work together. Furthermore, common technologies such as visualisation and decision support should be part of the platform. SOA is a technology that can be deployed to support this platform. According to the World Wide Web Consortium (W3C), SOA is 'a set of components which can be invoked, and whose interface descriptions can be published and discovered'. The CIE system architecture provided the common technology components and data access as services for each domain application.
- The business process drove the system architecture in Case Study 4. Different applications will be used at different stages of the process and the data requirements will be different. The process control mechanism was in place to ensure that the right data can be served to the right application.
- The services included technology service and data service. The technology service provided common technologies, such as visualisation of the building data, multi-aspects decision support, data analysis

for thermal, structural, conditions, etc. and process control mechanisms. The data service provided the data access to two types of data sources, building data and domain data. Building data could be described as data definition and representation of the building model and it was supported by interoperable data standards such as IFCs. Domain data was specific data related to each domain, such as regulations for building accessibility, weather data for energy simulation, etc.

References

Alshawi, M., Putra, C.W., Faraj, I., (1996), "A Structured Concept for Objects Life Cycle in Integrated Environments", Microcomputers in Civil Engineering, 12:339–351.

Aouad, G., (1997), "Integration: From a modelling dream into an implementation reality", Proceedings of CIB W55, International support for building economics, Lake District, England, pp. 7–29.

Aouad, G., Betts, M., Brandon, P., Brown, F., Child, T., Cooper, G., Ford, S., Kirkham, J., Oxman, R., Sarshar, M., Young, B., (1994), "ICON (Integration of Construction Information): Integrated Databases for the Design, Procurement and Management of Construction", Final Report, Department of Surveying and Information Technology Institute, University of Salford.

Aouad, G., Cooper, R., Fu, C., Lee, A., Ponting, A., Tah, J., Wu, S., (2005), "nD Modelling–A Driver or Enabler for Construction Improvement", RICS Research Paper Series, 5(6):1–16.

Aouad, G., Marir, F., Child, T., Brandon, P., Kawooya, A., (1997), "A Construction Integrated Databases–linking Design, Planning and Estimating", International Conference on Rehabilitation and Development of Civil Engineering Infrastructure Systems, American University of Beirut, Lebanon, Beirut.

Aouad, G., Sun, M., Bakis, N., Swan, W., (2001), "GALLICON Final Report", University of Salford.

Aouad, G., Wafai, M.H., (2002), "Implementation of Information Technology in the Construction Industry: The Conceptual Methodology Approach", 2nd International Postgraduate Research Conference In the Built and Human Environment, University of Salford.

Arayici, Y., Aouad, G., (2004), "DIVERCITY: Distributed Virtual Workspace for Enhancing Communication and Collaboration within the Construction Industry", The Proceeding of European Conference on Product and Process Modelling in the Building and Construction Industry (ECPPM), 8–10 September 2004, Istanbul, Turkey, pp. 415–422, ISBN 04 1535 938 4.

Aspin, R., Fernando, T., (2002), "Development and exploration of conceptual building function-to-form relationships", ECPPM 2002 European

Conference of Product and Process Modelling. eWork and eBusiness in AEC, Portoroz, Slovenia.

Augenbroe, G., (1995), "An Overview of the COMBINE Project, Proceedings". In R. Scherer, editor ECPPM'94: Product and Process Modelling in the Building Industry, Dresden, Germany, Balkema, pp. 547–554.

Brown, A., Cooper, G., Rezgui, Y., Brandon, P., Kirkham, J., (1996), "The architecture and implementation of a distributed computer integrated construction Environment", CIB Workshop: Construction on the Information Highway, Bled, Slovenia.

Cheng, T.F., (2005), "An IT Roadmap for Singapore's Construction Industry", Oslo, Norway, IAI Industry Day, May 2005.

Christianson, P., Da Dalto, L., Skjaerbaek, J.O., Soubra, S., Marache, M., (2002), "Virtual environments for the AEC sector", ECPPM Conference, Slovenia.

Cooper, R., Wootton, A., (1998), "Requirements Capture: Theory and Practice", Technovation, 18(8/9):487–511.

Coudret, F., Lombardo, J.C., Marache, M., Soubra, S., (2001), "A VR Application for the Construction Industry", Virtual Reality International Conference, Laval Virtual.

Da Dalto, L., Gobbetti, E., (2001), "The DIVERCITY project Deliverable 16&17: System Architecture, Multi-resolution Module and Communication Services", www.e-divercity.com.

Faraj, I., Alshawi, M., Aouad, G., Child, T., Underwood, J., (2000), "An IFC Web-Based Collaborative Construction Computer Environment: WISPER", Proceedings of the National Conference on Objects and Integration for AEC, Watford, United Kingdom, ISBN 186081 3771.

Federal Facilities Council, (2006), "A Cultural Revolution: Engineering, Construction, and Facilities Asset Management", Washington, DC, Federal Facilities Council – Government Industry Forum, http://www7.nationalacademies.org/ffc/government_industry_forum.html .

Fernando, T., Kähkönen, K., Leinonen, J., Murray, N., Tawfik, H., (2001), "Facilitation of collaborative communication for building construction with virtual reality technology", Conference on Applied Virtual Reality in Engineering and Construction Applications of Virtual Reality, Gothenburg, Sweden, pp. 1–17.

Fischer, M., Kam, C., (2002), "Product Model and Fourth Dimension (PM4D) Final Report", CIFE Technical Report No. 143, Center for Integrated Facility Engineering, Stanford California, 10/02, 2002.

Gobbetti, E., Spanò, L., Agus, M., (2002). "Hierarchical Higher Order Faces Cluster Radiosity", Technical Report No. CRS4 TR/, CRS4, Centre for Advanced Studies, Research, and Development in Sardinia, Cagliari, Italy.

Greening, R., Edwards, M., (1995), "ATLAS Implementation Scenario, Proceedings", ECPPM'94: Product and Process Modelling in the Building Industry, Dresden, Germany, Balkema, pp. 467–472.

Jordani, D., (2008), "BIM: A Healthy Disruption to a Fragmented and Broken Process, FAIA", Jordani Consulting Group, Journal of Building Information Modeling: Spring 2008, Houston, Texas, Matrix Group Publishing.

Kähkönen, K., (2001), "Visualising the Construction Progress", Industrial Horizons, VTT Customer Magazine.

Kam, C., Fischer, M., Hänninen, R., Karjalainen, A., Laitinen, J., (2003), "The Product Model and Fourth Dimension Project", ITcon, 8:137–166, Special Issue IFC–Product models for the AEC arena, http://www.itcon.org/2003/12.

Laiserin, J., (2003), "AEC Interoperability and the BLIS Project, CADALYST Magazine", www.cadalyst.com/cadalyst/.

Lee, A., Marshall-Ponting, A.J., Aouad, G., Wu, S., Koh, W.W.I., Fu, C., Cooper, R., Betts, M., Kagioglou, M., Fisher, M., (2003), "Developing A Vision of nD-enabled Construction", Construct IT, University of Salford.

Marache, M., Coudret, F., Lombardo, J., Aspin, R., Tawfik, H., Fernando, T., (2002), "Virtual Environments for the AEC Sector", eWork 2002, Prague, The Czech Republic.

Marir, F., Aouad, G., Cooper, G., (1998), "OSCONCAD: A Model-Based CAD System Integrated with Computer Applications", Electronic Journal of Construction Information Technology, 3:12–25 http://itcon.org/1998/3/.

Molina, J.M., Martinez, M., (2004), "XML Based Data Model for the Spanish AEC Sector", Proceeding of European Conference on Product and Process Modelling in the Building and Construction Industry (ECPPM), Istanbul, Turkey, pp. 149–154.

Nour, M., (2007), "Manipulating IFC Sub-models in Collaborative Teamwork Environments", ITC Digital Library, http://itc.scix.net/.

Patel, N., Campion, S., Fernando, T., (2002), "Evaluating the Use of Virtual Reality as a Tool for Briefing Clients in Architecture", Proceedings of the 6th International Conference on Information Visualisation, London.

Schevers, H., Mitchell, J., Akhurst, P., Marchant, D., Bull, S., McDonald, K., Drogemuller, R., Linning, C., (2007), "Towards Digital Facility Modelling For Sydney Opera House Using IFC and Semantic Web Technology", July 2007, http://itcon.org/2007/24/.

Shelbourn, M., Soubra, S., Martin, J., (1999), "DIVERCITY Design Review Workspace", DIVERCITY Project Deliverable 8.

Steinman, R., (2004), "MOBIKO, Mobile Cooperation in the Construction Industry Based on Wireless Technology", Proceedings of European Conference on Product and Process Modelling in the Building and Construction Industry (ECPPM), Istanbul, Turkey, pp. 521–526.

Sun, M., Aouad, G., (2000), "Integration technologies to support organisational changes in the construction industry", 7th ISPE International conference on Concurrent Engineering, Lyon, France, pp. 596–604.

Sun, M., Parand, F., (1998), "Integration of CAD, product model and distributed building component database", 2nd European Conference on Product and Process Modelling in the Building Industry, London, pp. 487–494.

Szigeti, F., Davis, G., (2003), "Portfolio and Asset Management: Performance Requirements", The IAI-NA PAMPer+ED Project, International Centre for Facilities.

Tawfik, H., Fernando, T., (2001), "A Simulation Environment for Construction Site Planning", The 5th IEEE, International Information Visualisation Conference, Southbank University, London.

Thery, S., Da Dalto, L., Dupuy, Y., Balet, O., Paulin, M., (2001), "Realistic Rendering in Interactive Architectural Design", VRIC, Virtual Reality International Conference, Laval Virtual.

Zarli, A., Richaud, O., (1999), "Requirements and Technology Integration for IT-Based Business-Oriented Frameworks In Building and Construction Journal", ITcon, 4:53–75, http://itcon.org/1999/4/.

4

Requirements Engineering in CIE Development for the Construction Industry

4.1 Introduction

The effective uptake of highly advanced technology is not a straightforward process. In the first instance, it might seem that large investments in IT could bring about benefits to its investors, which is actually not always true. On the contrary, previous experience and studies have shown that the technology-push system development is not sufficient to improve the efficiency and effectiveness of work environments without clear consideration of the business processes and user-related issues.

Process improvement and modelling have been associated with a number of ideas concerned with the dynamic behaviour of organisations, businesses or systems in a more general manner. Process models constructed from the users, or stakeholders, viewpoints can form the basis for computer systems used to support particular behaviours for an organisation (Snowdon, 1998). Therefore, process-based computer integrated environment (CIE) systems are more likely to be successful in meeting the organisation's needs and requirements and are able to introduce overall improvements to the business environment (Alshawi and Faraj, 2000).

The approach of CIE development has been technologically driven. 'Push strategy' is the most dominant pattern of IT procurement in the construction industry. This has led to a large number of failures being reported in utilising CIE due to the insensitivity of these systems to organisational design, external environment and culture and management system. Therefore, there has always been a gap between the underlying philosophy of IT and the environment in which it is implemented (Aouad and Wafai, 2002). Strategic exploitation of IT in the construction industry will be achieved by transferring these solutions into practical uses, based on social process and business strategic points of view, not just from merely a technological point of view. Requirements engineering (RE) is the discipline that brings technology, people and process together for the development and implementation of CIE.

Failing to strategically utilise and implement the advanced technological solutions in the construction industry could be attributed to the fact that the same techniques and methods, which were developed in other industries, have been applied without enough studies of their fit for the construction industry. The implementation of CIE technologies should, therefore, follow an agreed methodological plan, which considers the different dimensions of construction organisations for implementing and developing CIE technologies.

4.2 CIE Systems from Technological Perspective

CIE systems are criticised by researchers from the technological perspective and the following drawbacks are experienced in the development and implantation of CIEs (Zarli and Richaud, 1999; Arayici, 2004):

- **Homogeneity:** Solutions are fixed and not open, with a lack of support for legacy, as well as new, upcoming systems in terms of hardware, software, databases and networks.
- **High entry level:** Solutions are often too expensive to employ for small and medium size enterprises (SMEs). There have to be more entry levels, e.g. cheap personal to costly enterprise editions.
- **Lack of scalability:** Limited growth path in terms of hardware and software usage.
- **Application centricity:** Need to organise the enterprise around the application.
- **Fixed infrastructure:** Need for leased lines among partners, restricting location independence and requiring long-term relationships.
- **Lack of support for business processes:** Limited security and transactional support.

The following is required to enable CIE to provide the ability to transact business processes seamlessly and to capture and access and assess the state of the business/project and industry needs (Zarli and Richaud, 1999; Arayici, 2004):

- Low-level entry
- Scalability
- Open infrastructure and location-independent access
- Enterprise information (i.e. seamless capturing of the state of business from distributed legacy data)

- Support for business processes
- Security and transactional support.

Commercial pressures often dictate that the existing construction industry software applications satisfy a very broad range of user needs. This results in monolithic applications. Component-based applications, or service-based applications, could be assembled on the basis that user needs and requirements could be more portable, thereby creating the opportunity to involve the construction industry practitioners, who have not been included effectively, in the CIE system development. This may promote the use of component-based CIE or service-based CIE on site. Component-based and/or service-based applications are also easier to distribute over networks.

4.3 Requirements Engineering in the CIE Community

Process-model methodologies for system development such as the code-and-fix model, the stage-wise model, the waterfall model, the evolutionary model, the transform models, Boehm's spiral model, the rational approach, the agile processes, Extreme Programming (XP) techniques, etc. have been evolving since computer technologies started to be used (Bohemn, 1988). It is stated that throughout the period from the code-and-fix model to the agile processes and XP, the awareness of the computer society about RE has gradually improved. In other words, whilst the user requirements were paid hardly any attention and effort in the early software development processes, it is contemporarily recognised as an engineering domain in the state-of-the-art development processes. A similar evolution about user requirements in the CIE development is also taking place specifically for CIE developments.

The emerging technologies of CIEs and concurrent engineering for architecture, engineering and construction (AEC) enables the widespread communication and collaborative use of project information among all project participants and all project life cycle stages using computing technologies. Early research prototypes and process models for CIEs for construction explored the possibility of such information integration to create a lean construction environment and eliminate the drawbacks highlighted in Chapter 3.

Over the past few years, a number of significant CIE projects have been carried out that incorporate some type of AEC process models. Some early versions of the CIE process models such as Building Project Model (Luiten, 1994), Unified Approach Model (Bjork, 1992), The General Construction Object Model (GenCOM) (Froese, 1992),

Architecture, Methodology and Tools for Computer Integrated Large-Scale engineering (ATLAS) Project type model (Tolman *et al.*, 1994) and Information/Integration for Construction (ICON) (Aouad *et al.*, 1994) were the initial contributors to the concept in terms of idea generation. In order to point out the evolving maturity and the awareness of the CIE community about RE in each development, some examples of the CIE systems are introduced in Chapter 3 and this chapter will elaborate on these developments with respect to the RE.

It will be realised that while there was hardly any awareness about RE in the early developments, which were actually investigating the possibility of such a CIE concept, in the later ones there was an increasing awareness about RE. The main concern has been gradually diverting from searching for the possibility of such CIE for the construction industry to its adaptation and implementation in the industry.

4.3.1 The ATLAS system

ATLAS is a collaborative European project that aimed to improve communications in construction projects by using information exchange standards such as Standard for the Exchange of Product Model Data (STEP). The ATLAS research team mainly investigated the integration, data exchange and communication strategies using the STEP data model. Furthermore, there was an attempt to employ knowledge-based and case-based systems to support specific business functions and effect a major role with respect to the development of versatile, easy-to-use (easy-to-modify) rule-based converters and semantic models (Poyet, 1995).

Although it was a prototype development for the construction industry, the project team did not consider RE at all in the development life cycle. This is because there was no awareness or consideration for the implementation of the system in ATLAS development. Although the ATLAS system helped to prove the possible use of CIE in the construction industry, it was not a user-driven system. This is because RE was not included in the project, although the IDEF0 methodology was used for functional and business process analysis (Poyet, 1995).

4.3.2 The OSCON system

The Open Systems for Construction (OSCON) system employed the process model of the ICON approach (Aouad, 1997), which is a top-down approach based on hybrid information and an object-oriented approach to establish the principles for integration. As a result, the first goal in the OSCON project was to achieve information integration, which resulted in the lack of appreciation of RE.

During the OSCON development, the main concern of the construction CIE community was the product modelling issues. Modelling design information is the most debatable aspect of designing the product model. It involves dealing with graphical and non-graphical data and many complex issues, such as representing form, function, topology, appearance and many other aspects (Marir *et al.*, 1998).

Several research studies suggested models for space and space enclosure and also developed object-oriented product models that could be used to store richer kinds of data and knowledge about a product, including the knowledge about its design, manufacturing and operational parameters. However, none of these models has had universal acceptance. This is due to the fact that a data model must be readily accessible to all construction-related applications and users. In view of this, a group of software companies and users came together to form the International Alliance for Interoperability (IAI) and define specifications for a set of Industry Foundation Classes (IFC).

During the OSCON development, the concept of CIE was still immature, but improving in terms of integration. It was becoming sophisticated and comprehensive while some integration aspects, such as IFCs, which will help in the development of CIEs for construction, were still emerging. As a result, RE did not have any priority at all in the OSCON system development, because the other issues had priorities such as process models, proving the efficacy of the CIE concept, etc. Although class and object diagrams were developed for the architectural design of the OSCON system, use case models were not developed for requirements modelling. In fact, software specifications are derived and class diagrams should normally be developed from use case models, which did not occur in OSCON.

In conclusion, there was no effort invested on RE in the OSCON system development. Therefore, it is classified as a technology-oriented system. It could have been a successful development to demonstrate the CIE concept; however, it was a failure because it was not a practical system for the industry; the failure could be attributed to the missing process, people and cultural issues, which could have been bonded through RE during the development.

4.3.3 The SPACE system

SPACE stands for Simultaneous Prototyping for an Integrated Construction Environment. It employed a project model rather than a process model. SPACE framework was established to coordinate the integrated process between various construction applications. The system can provide integration of construction information and concurrent work and information exchange (Faraj and Alshawi, 1999).

In respect to RE, the SPACE prototype was tested with a number of cases to validate the implementation of the system. However, these tests were not considered as a way of capturing user requirements to improve the system accordingly. They were only conducted for the demonstration of the system to the industrialists at the end of the project life cycle.

There was no effort spent on the user requirements capture and modelling in the SPACE system development. It is a highly automated system and as a result, it was a powerful solution from a technology perspective. However, it did not fit the construction industry and it was technology oriented as opposed to being user oriented. That is to say, if it was considered to be adapted to the industry it would not have been possible.

4.3.4 The WISPER system

WISPER is the acronym for Web-based IFC Shared Project EnviRonment. Many emerging technologies were brought together, such as IFC, Web technology, Common Object Request Broker Architecture (CORBA) and Virtual Reality Modelling Language (VRML). Standard off-the-shelf software was also used (Faraj *et al.*, 2000). The architecture of WISPER uses a three-tier client-server infrastructure to demonstrate the integration among detailed design, building element-based cost estimating, construction scheduling and a VRML viewer that allows for graphical querying of a project database (Alshawi *et al.*, 1998). Within a three-tier architecture, each tier/facet of the application – i.e. presentation (user interface), logic (processing), data storage (database) – was isolated, providing benefits such as maximum control, scalability and flexibility.

Compared to previous CIE solutions, in the WISPER development there was an awareness of RE. Prior to the implementation, the research team set out to investigate the requirements from the end user's point of view. The use case technique was selected for RE (Faraj and Alshawi, 1999). However, use cases were not adequate to capture all the user requirements.

The use case technique is not sufficient for RE, as use case models are a loose collection of use cases. They are not suitable for requirements elicitation. They mainly address the functional requirements, and are not scalable, which means that they can be either of very high or of very low level; if it is a high level use case, it can be understandable but not useful for the developers.

On the other hand, if they are of a low level, they can be understandable and useful to the developers, but the end users would struggle to understand them. They do not represent the system work model well. In other words, use case definition does not indicate precisely where a use case starts and where it ends (Beyer, Holtzblatt, 1998; Robertson and Robertson, 1999).

Consequently, they need to be supported by additional techniques to properly synthesize the use cases for cohesive RE. However, in the WISPER project, the only technique used in the project was the use case approach, which is considered insufficient for the collaborative, distributed, integrated information systems such as WISPER.

The use cases included in Table 4.1 were developed in the WISPER system (Faraj *et al.*, 2000).

Although RE was inadequate and not properly done in the WISPER project, it was promising to see RE activities in WISPER development as it shows that there was an emerging awareness about them in the CIE development.

4.3.5 The GALLICON system

The main purpose of the GALLICON project was to attempt to extend this into a real-world environment, supporting live processes (Aouad *et al.*, 2001). The system was developed with the industrial users providing information for the process modelling. In the GALLICON project, there was more awareness of RE than any other previous CIE projects.

Interviews were conducted with the industrialist participants to assess the information flow between the project partners. It was a

Table 4.1 Main use cases identified in the WISPER project.

Use Case 1	Establishing a new project database
Use Case 2	Populating a database with design information
Use Case 3	Generating cost estimates
Use Case 4	Generating specification of building elements
Use Case 5	Linking specifications to individual building elements
Use Case 6	Browsing data through a VRML viewer
Use Case 7	Generating construction-planning information
Use Case 8	Accessing data sheets

process-led development; however, the number of interviews was limited to the project participants. Yet, holding workshops and special interest groups and interaction and communication with end users did not occur and a vision definition, on which there would have been a common consensus between the end users, did not materialise. Lastly, there was no clearly identified formal RE process in the project.

The GALLICON system infrastructure was developed while the process modelling was carried out, which may project the XP technique as an agile RE technique. However, it was not done with an awareness of RE. Besides, a missing symbiotic relationship with the end users could cause misunderstandings, misinterpretations and mixing the priorities of the end user issues. Furthermore, the software infrastructure may not reflect the right system architecture to satisfy the user requirements. Part of the challenge stems from the fact that requirements and architectures leverage different terms and concepts to capture the artefacts relevant to each (Grunbacher *et al.*, 2001). However, the GALLICON project had weaknesses in interpretation and transition from the requirements to the system design. Subsequently, it was open to the threads coming from sequential process models for the development of the system architecture (Beyer and Holtzblatt, 1998).

The end users may not fully conceive the defined software infrastructure, which can hide important drawbacks. The process models are sequential and include threads, and are understandable to the end users. System design is structural and it should be able to reveal how all the threads fit together coherently, but it is not easy to be conceived by the end users (Beyer and Holtzblatt, 1998); therefore, the transition from sequential thinking to structural thinking is crucial, but it was weak in the GALLICON project. Lastly, requirements management and validation hardly existed in GALLICON.

4.3.6 The DIVERCITY system

Distributed Virtual Workspace for Enhancing Communication within the Construction Industry (DIVERCITY) aimed at supplying a shared virtual construction design and briefing environment that enables the construction stakeholders to conduct client briefings, design reviews, simulate 'what if' scenarios, test the constructability of buildings and communicate and coordinate design activities between teams.

The project team considered that because the construction industry was a large multi-disciplinary industry, with multi-faceted perspectives and requirements, DIVERCITY needed to define broad industrial requirements and expand into more detail to capture the briefing and design requirements.

DIVERCITY reviewed a number of methodologies and finally used a combination of techniques. The DIVERCITY requirements approach is relatively mature, compared to the other CIE system developments. It was divided into three categories: (i) the contextual design, (ii) the use case driven requirements analysis, storyboarding and (iii) the incremental prototyping with user tests, each of which was a complementary strong technique to undertake the aspects of requirements elicitation, requirements analysis and specification and requirements validation respectively.

There were not so many well-formalised methods to support the entire design of a product like DIVERCITY. Owing to its well worked-out user-centred approach, the contextual design method (Beyer and Holtzblatt, 1998) was chosen to take into account the end user's work practice and interface requirements. Furthermore, because system developments start with a Unified Modelling Language (UML)-based object modelling with efficient tools like Rational Rose from Rational Software Corporation (http://www.rational.com) it takes very little consideration of the early conceptual analyses of user requirements and user interface functionality (Christiansson *et al.*, 2001). The risk was obvious that important properties of the final product with regard to the end user are overlooked. Besides, as previously mentioned use case models are a loose collection of use cases. Hence, employing the contextual design before use case modelling would bring about a strong synthesis through which separate use cases are integrated well.

The DIVERCITY design process involved the end users in an incremental prototyping process as part of the RE process. It was extremely important to bridge the gap between the user requirements specifications and the actual interface design and the implementation of the underlying operational models of the distributed virtual workspace system (Christiansson *et al.*, 2001). Although DIVERCITY employed the state-of-the-art techniques for the user requirements process, there have been some difficulties in the development life cycle of the project. This was due to a lack of communication between the user groups and the developers, and a lack of user requirements strategy or requirements management and shared understanding.

4.3.7 The nD modelling system

Compared to the previously mentioned systems, in terms of a shared understanding and vision between both practitioners and researchers, the nD modelling development is considered to be the most mature and popular development in CIEs. A series of workshops were conducted to gain a consensus on the boundaries of the nD modelling concept. The workshops were used to define, develop and validate the nD modelling system and included an academic research team workshop and a national and international workshop.

The first academic workshop set about defining the need and scope of nD modelling using a case study exemplar. To identify the similarities and differences in design requirements between various disciplines, a provisional plan of an office space for 10 researchers was circulated with instructions that improvements and developments should be made to the design, based on individuals' disciplines such as client, end user, accessibility, sustainability, maintainability, security, acoustics and energy (Aouad *et al.*, 2005).

The purpose of the national workshop was to gain support for nD modelling and to ensure that these were in line with the demands and requirements of the United Kingdom's wider academic and industrial communities. The attendees included contractors, clients, suppliers and architects, ranging from one-person organisations to large multinational organisations.

The spectrum of participants was to reduce any inherent exclusion of industry players in the development of nD and to gain interest in and acceptance of the work. The aims were to be achieved by asking the group to discuss two questions: 'How may IT support integrated design and construction in the future, in both 5 and 20 years time'; and 'What are the barriers and opportunities for implementation?' (Aouad *et al.*, 2005).

Finally, the international workshop brought together leading players from around the world, again from both the industry and academia. It was envisaged that this event would augment the findings from the previous workshops. The workshops of most relevance to the question of the practicality of the nD modelling vision's propositions were the national and international workshops as these are focused, not upon the vision of the 'ideal' construction industry, but upon the practicality of that vision's propositions in dealing with the problems and issues that the industry stakeholders actually experience, including the opportunities and barriers to the implementation of nD modelling (Aouad *et al.*, 2005).

Workshops played a big role and made an impact in identifying a shared vision for the nD modelling concept, to capture the user requirements and to specify the scope and limitations of the nD modelling system. Although workshops can help for requirements engineering, there was no specific formal RE process that embraced consecutive complementary techniques for requirements elicitation, modelling, specification, validation and requirements management. On the other hand, it is noteworthy that the main concern in the nD modelling development opportunities was the barriers put up by the industry to the uptake of the system, whereas earlier development had focused on the possibility of using the CIE concept for the construction industry. This proves that there has been an evolvement of CIE over time.

These developments also showed that there are barriers against the implementation of CIE in the construction industry, such as cultural, process, legal, educational and technological issues. However, these issues can be considered and modelled into the specification of the CIE developments via correct RE processes, which can contribute to the successful development and implementation of CIE systems.

4.4 Interviews in the Construction CIE Community

A number of interviews were conducted with academics and industrialists about RE in CIE development. The purpose of the interviews was, to an extent, a critical analysis in order to support the justification of the need for RE in CIE development. The interviewees were chosen from a variety of professions. However, their profession and expertise are related only to the construction industry and IT. The findings from the interviews are analysed and presented in Figure 4.1. The interview findings were clustered in six categories. These are (i) importance of RE in CIE development, (ii) influence of RE upon the implementation, (iii) lack of RE in CIE developments, (iv) increasing awareness about RE in the CIE community, (v) the main criteria for the RE approaches and (vi) evaluation of the RE approaches. The map in Figure 4.1 illustrates the concepts and their associations and defines a preliminary scope for the interview findings.

4.4.1 Importance of requirements engineering in computer integrated construction (CIC) development

It is needless to say that RE is crucial in CIE development for the construction industry.

The difficulty is how to validate that the captured requirements are in fact the correct ones. This is because, if the requirements are not captured correctly, some user issues cannot be reflected on to the system. In other words, the system can demonstrate some functionality that is irrelevant to the user's needs. However, any software that is developed

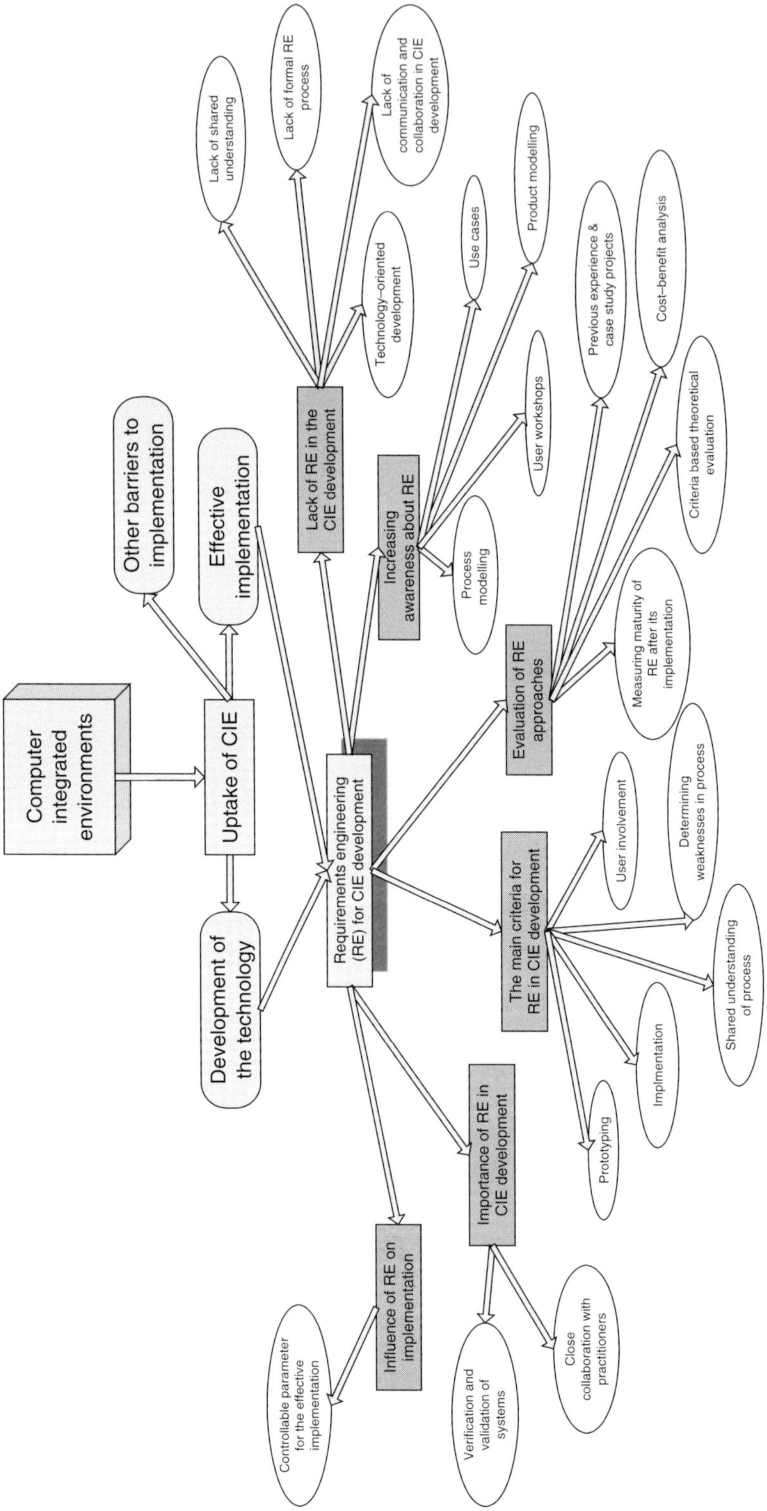

Figure 4.1 The concept map of the interview findings about RE in the construction CIE community.

for the construction industry has to be focused on the end users. There is no point in developing software for any domain that is not focused on what the end users want. Otherwise, it will result in an obvious failure of the system and affect the implementation by the users negatively.

Besides, the construction industry is not knowledgeable about IT systems, in particular CIE systems. Therefore, there is a certain amount of leading required, along with RE in the CIE development. It is not sufficient to allow the construction people to define the requirements because they do not have enough understanding to give the developers what they really need. It should be overcome through RE. However, RE was not considered enough by the construction CIE community. Furthermore, the correct requirements must be captured using the right techniques. If requirements are not captured correctly, the system cannot provide the right functionality for user demands. The difficulty is how to conduct the user requirements activities effectively, which will enable the CIE developers to capture the right requirements and in the right way.

On the other hand, software developers and vendors are focusing on certain parts of the construction projects, such as planning, cost management, project management and computer-aided design (CAD), which has led to the island of automation. However, the integration was not considered favourably until building information modelling (BIM) became popular recently. This is because the value and importance given to lean construction via CIE was not properly understood by the developers whilst the construction practitioners did not understand the RE issues in system development. As a result, practitioners do not know how to inform the developers about their need to develop the systems accordingly. Consequently, it is vital to bridge the gap and identify a strategy for RE for CIE systems.

4.4.2 Influence of requirements engineering upon implementation

RE has an influence on the implementation of CIE and could make a certain contribution to the same. However, it cannot overcome all the problems relevant to the implementation of CIEs because RE is not the only parameter that influences it.

There are several other aspects that also affect the implementation, such as cultural issues, legal issues, traditional procurement philosophies, technological issues, contractual issues, processes and educational issues, etc. These parameters can be classified as uncontrollable parameters on implementation, whereas RE can be classified as a controllable parameter for implementation. In other words, better RE in the development can help to improve the implementation of the CIE system.

The CIE systems that were developed for the construction industry are very much product-oriented systems. However, the construction industry is a service-oriented industry. Therefore, the systems should be service-oriented, not product-oriented, for the successful development and implementation in the industry. In addition, rather than a sequential process, a symbiotic relationship should be developed between RE and implementation teams, which should occur during the phase of the project when early prototypes are delivered.

The current BIM trend openly shows that the CIE systems will be of use in the construction industry in the future, but the industry is still struggling to differentiate itself from its conventional structure towards lean process-based structure with CIE support. When BIM concept has a better and wider acceptance and adaptation, the influence of RE in the CIE development will subsequently be much higher than it is today.

4.4.3 Lack of requirements engineering in the CIE developments

Most developments (i.e. research prototypes) in the last two decades are the proof of the CIE concept from a technology point of view. Very few of them, such as the DIVERCITY project, included RE. In the early prototype developments, RE was not taken into account. However, RE is an important part of the CIE development, and even the development of data models such as IFC incorporate intensive requirements activities.

Projects like WISPER, GALLICON and DIVERCITY are examples in which there is evidence of RE being taken into account although it was done in an ad hoc manner. For example, in the WISPER projects, use cases were developed. This helped the industrialists to understand the functional behaviour of the system.

In the GALLICON project, RE was modelled through scenario modelling, while in DIVERCITY, RE was modelled through use case modelling, contextual design and prototyping. RE enabled these systems to be practical, usable and adaptable, configurable and scalable CIE systems even though the modelling was not done correctly.

On the other hand, CIE developments such as ATLAS, OSCON and SPACE systems were not user oriented; they were purely technology oriented. In these developments, there were other crucial priorities and there was no awareness about RE. Therefore, they were not practical and adaptable systems, unlike the GALLICON and DIVERCITY systems. However, this does not mean that RE was sufficient, mature and systematic in those projects such as DIVERCITY.

Although the most sophisticated formal RE amongst these projects was done in DIVERCITY, there was still immaturity in the DIVERCITY RE such as lack of requirements management, lack of shared and common understanding, lack of communication to the stakeholders and so on.

Last but not the least, current IT tools such as AutoCAD are production-oriented developments, which cannot handle complex, topological data sets such as IFC. On the contrary, new-generation toolkits are targeting the design processes of construction from early design to detailed design, construction and facilities management, consequently highlighting the importance that must be given to RE. Also, there is a great need to understand the connections to a larger context, where the end user's value chain requirements and procurement systems' demands are the driving factors, i.e. research efforts should be driven by end users' needs rather than information and communication technology (ICT) solutions (Arayici *et al.*, 2006; Nour, 2007).

4.4.4 Increasing awareness about requirements engineering in the CIE community

BIM is an evolving CIE concept in the construction industry as a BIM and subsequently the subject of RE in the CIE development is also a slowly evolving issue. Furthermore, research studies to define the best RE approaches and methodologies for system and data model development have been undertaken already.

From a researchers' point of view, there is an evolving awareness about RE. Academics have realised the importance of RE. On the whole, in the construction CIE community including academia and practitioners, the awareness is very less and is evolving very slowly. The evidence of this awareness is the use of RE techniques in the latest CIE developments such as use case modelling. For example, the WISPER and GALLICON systems were more practical and user friendly than SPACE and OSCON, because the WISPER and GALLICON systems were user oriented compared to the SPACE and OSCON systems.

In the DIVERCITY example, there was a balance between the end users and developers. The end users found it usable and acceptable and the developers found it achievable and client focused. At the beginning of the project, the users group and the developers worked separately without interacting and communicating. Six months from the outset, the project consortium realised that this was not working and there was a need to tie up relationships between the user groups and the technology group so that the end users were interacting directly with the technology group.

The work of the user groups in the first 6 months in the DIVERCITY project implied gathering generic requirements without focusing on the technology available, which did not work. Likewise, developing the technology without focus on the requirements also did not work. Therefore, a relationship was established between the two groups, which provided a strong development platform.

Another drawback was that as the end users, both in the DIVERCITY project and in general construction, stakeholders are not really aware of the technology available from which they can benefit. Therefore, education and training are necessary for the end users to increase their awareness about the technology available. Subsequently, the requirements can be updated properly and a more rational requirements specification can be provided to the developers.

On the other hand, the construction industry is fixed in what they have traditionally. On the whole, they are predominantly trying to re-implement, with IT support, what they have done as a manual process, which means that they take manual processes and translate them into computer-based processes and IT-supported processes. What they do not do is look at what IT systems can do for them and how IT systems can affect the way they are working.

4.4.5 Main criteria for requirements engineering activities

The most important criteria are the users and their involvement in the CIE developments. They provide the relevant data to form the requirements specifications for the system development. For example, use case analysis from UML is a good tool to formalise the user requirements. However, it is not sufficient because use cases help us to capture only the functional requirements, and it is not scalable.

Furthermore, prototyping is also another important criterion, which should be taken into account in RE. This is because the users want to see how the system runs and to evaluate it accordingly in different respects such as functionality, usability or practicality and also ease of use and other quality aspects. Thus, early prototypes should be developed and shown to the users by doing black-box tests. The construction users would like to see the actual system. That enables the developers and researchers to incrementally develop the system based on the user's feedback and criticisms. In addition, (i) understanding what the process is and (ii) determining the weaknesses in the process with end users are also highlighted by the interviewees as criteria for selecting and adopting the RE techniques.

As mentioned previously, the construction industry is fragmented and has a lot of best practices and advisory documents, but it does not have any defined processes. Therefore, the first thing to do in conducting requirements elicitation is to define what the process is from the point of view of the developers. Once the shared understanding of the process is clear, then the core user requirements are defined accordingly. For example, in the DIVERCITY project, the project team has defined the process in which the understanding of the developers and users are mapped. Process definition needs to be done collaboratively between the developers and user groups.

Another example is the GALLICON project, which also identified a process for the targeted construction activities. Process Protocol Map (Kagioglou *et al.*, 2000) itself was a project to specify a lean construction life cycle process. Next, once the process is defined from the developers' understanding, then determination of weaknesses in the process should be done, and through interaction with the end users, these weaknesses are addressed.

RE must be done through two-way messaging between the user groups who define what they want to do and the developers who decide what they can do with the technology available. Therefore, there should be two-way relationships going on. Lastly, implementation is the ultimate criteria for the evaluation of RE approaches employed in the system development because RE affects the uptake of the systems. However, RE was not considered sufficiently by the CIE community for the development of the CIE technology and its effective implementation.

4.4.6 Evaluation of the requirements engineering approaches

It was also asserted from the interviews that the construction CIE community should try techniques until they see which works well enough to carry out the requirement engineering. In addition, previous experiences and lessons learnt from previous cases can also be used to evaluate the RE techniques. A mechanism to try out the techniques to quickly decide on the right one does not exist.

Return-on-investment analysis should be included in RE by doing cost–benefit analysis to justify the system development and the implementation of the system in the industry. However, cost–benefit analysis did not exist in any of the CIE development.

The use of a number of criteria that reflect the characteristics of the domain area is possible in order to decide the most appropriate techniques for RE. Furthermore, an implemented RE process can be assessed to measure its maturity by means of the capability maturity models and improve it accordingly.

However, no scientific evaluation and research on RE in the CIE development and the relevant criteria within the RE approach has been done. In this case, defining a comprehensive and complete RE process would be useful to the CIE system development.

References

Alshawi, M., Aouad, G., Faraj, I., Child, T., Underwood, J., (1998), "The implementation of the Industry Foundation Classes in integrated environments", CIB W78 Conference, August, 1998, Sweden, pp. 55–66.

Alshawi, M., Faraj, I., (2000), "Integrated Construction Environments: Technology and Implementation", Construction Innovation, 2(1):33–51.

Aouad, G., (1997), "Integration: from a modelling dream into an implementation reality", Proceedings of CIB W55, International Support for Building Economics, Lake District, United Kingdom, pp. 7–29.

Aouad, G., Betts, M., Brandon, P., Brown, F., Child, T., Cooper, G., Ford, S., Kirkham, J., Oxman, R., Sarshar, M., Young, B., (1994), "ICON (Integration of Construction Information): Integrated Databases for the Design, Procurement and Management of Construction", Final Report, Department of Surveying & Information Technology Institute, University of Salford.

Aouad, G., Cooper, R., Fu, C., Lee, A., Ponting, A., Tah, J., Wu, S., (2005), "nD Modelling–A Driver or Enabler for Construction Improvement", RICS Research Paper Series, 5(6):1–16.

Aouad, G., Sun, M., Bakis, N., Swan, W., (2001), "GALLICON Final Report", University of Salford.

Aouad, G., Wafai, M.H., (2002), "Implementation of Information Technology in The Construction Industry: The conceptual Methodology Approach", 2nd International Postgraduate Research Conference in the Built and Human Environment, University of Salford, Greater Manchester, United Kingdom.

Arayici, Y., (2004), "Requirements Engineering In Innovative System Development For the Construction Industry2", PhD Thesis, The University of Salford.

Arayici, Y., Ahmed, V., Aouad, G., (2006), "A Requirements Engineering Framework for Integrated Systems Development for the Construction

Industry", Electronic Journal of Information Technology in Construction, III: 35–55, http://itcon.org/2006/3/.

Beyer, H., Holtzblatt, K., (1998), "Contextual Design. Defining Customer-Centered Systems", San Francisco, California, Morgan Kaufmann Publishers.

Bjork, B.C., (1992), "A Unified Approach for Modelling Construction Information", Building and Environment, 27(2):173–194.

Bohemn, B.W., (1988), "A Spiral Model of Software Development and Enhancement", TRW Defense Systems Group, 0018-9162/88/0500-0061.

Christiansson, P., Svidt, K., Skjærbæk, J.O., Aaholm, R., (2001), "User requirements modelling in design of collaborative virtual reality design systems", International Conference on Construction Information Technology, Mpumalanga, South Africa.

Faraj, I., Alshawi, M., (1999), "A Modularised Integrated Computer Environment for the Construction Industry: SPACE, Salford, Lancashire, University of Salford, http://itcon.org/1993/3/.

Faraj, I., Alshawi, M., Aouad, G., Child, T., Underwood, J., (2000), "An IFC Web-Based Collaborative Construction Computer Environment: WISPER", Proceedings of the National Conference on Objects and Integration for AEC, UK, ISBN: 186081 3771.

Froese, T., (1992), "Integrated Computer Aided Project management Trough Object Oriented Models", PhD thesis, Department of Civil Engineering, Stanford University.

Grunbacher, P., Egyed, A., Medvidovic, N., (2001), "Reconciling Software Requirements and Architectures: The CBSP Approach", Fifth International Symposium on Requirements Engineering, Los Alamitos, California, IEEE. Computer Society Press, pp. 202–211.

Kagioglou, M., Cooper, R., Aouad, G., (2000), "Performance Management in Construction: A Conceptual Framework", The Journal of Construction Management and Economics, 19(1):85–95.

Luiten, G., (1994), "Computer Integrated Design for Construction in the Building Industry", PhD thesis, Faculty of Civil Engineering, Delft University of Technology, The Netherlands.

Marir, F., Aouad, G., Cooper, G., (1998), "OSCONCAD: A Model-Based CAD System Integrated With Computer Applications", Electronic Journal of Construction Information Technology, 3:12–25, http://itcon.org/1998/3/.

Nour, M., (2007), "Manipulating IFC sub-models in collaborative teamwork environments", Proceedings of the 24th CIB W-78 Conference, Maribor, Slovenia, 26-29, ISBN 978-961-248-033-2.

Poyet, P., (1995), "ATLAS integration tools", Proceedings of the 1st ECPPM Conference, Dresden, Germany.

Robertson, S., Robertson, J., (1999), "Mastering Requirements Engineering Process", Boston, Massachusetts, ACM Press, Addison Wesley, ISBN 0201360462.

Snowdon, R.A., (1998), "Overview of Process Modelling", http:/www.cs.man.ac.uk/ipg/Docs/pmover.html.

Tolman, F., Bakkeren, W., Bohms, M., (1994), "ATLAS LSE Project Type Model", ASPRIT Project 7280 – ATLAS/WP1/Task 1500 Document D106, www.newcastle.research.ec.org/esp-syn/text/7280.html.

Zarli, A., Richaud, O., (1999), "Requirements and Technology Integration for IT-Based Business-Oriented Frameworks in Building and Construction", ITcon, 4:53–75, http://itcon.org/1999/4/.

5

Evaluation of Requirements Engineering Processes

5.1 Introduction

Central to an understanding and improvement of the requirements engineering (RE) process is the ability to measure RE success (Emam and Madhavji, 1995). Existing claims (Brooks, 1987; Davis, 1989) and empirical evidence (Wu, 1990; Keil *et al.*, 1998; Carr 2000; Nikula and Sajaniemi, 2002) support the notion that an inadequately performed RE process is positively associated with system failure. Thus, there is an economic and software quality payoff in improving the RE practices. However, a prerequisite to improving RE practices is a fundamental understanding of RE processes and the parameters that lead to their success, or cause their failure.

RE techniques have been assessed and evaluated using case studies, pilot projects and experiments (Porter and Votta, 1998; Cooper and Ito, 2000). The evaluations are used to discover the strengths and weaknesses in a particular technique, to determine the costs of applying the technique, or to determine the benefits of using the technique. The evaluation provides qualitative and/or quantitative results, which can be used to decide if the technique is going to be used on a particular project.

In this chapter, RE practices are elaborated on from various points of view in order to pick up the key issues from discrete research studies. These key issues, extracted from each perspective, are discussed in terms of their suitability for computer integrated environment (CIE) systems. Also some issues such as the gap between the research and industrial uptake are interrogated. Furthermore, a comprehensive research study about measuring the success of RE processes through a set of evaluation criteria is introduced. Finally, the key issues and the criteria are comparatively analysed and evaluated in order to confirm the validity of the criteria.

5.2 Improving the Requirements Engineering Process

The software systems of today are characterised by increasing size, complexity, distribution, heterogeneity and lifespan. They demand careful capture and modelling of requirements and architectural designs at the earliest stages, before the low-level system details begin to dominate the engineers' attention and significant resources are exhausted for system construction. Researchers and engineers have approached requirements issues from different perspectives to enhance the RE processes (Arayici, 2004; Arayici *et al.*, 2006). The following sections briefly address these perspectives and some evaluation criteria from these perspectives are observed.

5.2.1 Traceability through product and process modelling

Industrial software intensive systems are affected by unavoidable requirement changes that can occur during any phase of the production cycle and have strong effects on a variety of technical and managerial aspects. Furthermore, traceability (Lavazza and Valetto, 2000) provides a means to analyse and keep track of the consequences of requirements changes over the entire software product.

It is recognised that the amount of knowledge captured by means of traceability can be invaluable for a software development organisation, but that knowledge often remains implicit and can be scarcely exploited without proper formalisation and structuring. Added to that, traceability must be customised according to the specific organisation as well as the project-specific goals to maximise its usefulness (Domges and Pohl, 1998).

Traceability-based methodologies are proposed by researchers to improve the effectiveness of requirements management. These methodologies were implemented in different case studies in order to assess their quality in handling the complexities that are inherent in the technical, managerial and contractual dimensions of requirements and change management (Lavazza and Valetto, 2000).

Two main dimensions are determined in these methodologies. These are (i) the process of RE and (ii) the product of the system development. These two dimensions are measured using an appropriate measurement approach such as the Goal/Question/Metric (GQM) technique (Basili and Rombach, 1988), which forms a basis for the quantitative analysis and quantitative assessment of the impact of requirements changes and quantitative estimation of cost of the development activities.

Requirements management is conducted with process and product awareness in mind, and requirements traceability is modelled according to the product and process. The conceptual models of the product and process can be exploited to identify, organise, situate and maintain various kinds of traceability relations. Such situated traces can either originate from semantic inter-artefact relationships explicated by the product model or from the data flow described by the process model. The product-based traceability spans from the user change request and system requirements (pre-requirements traceability) to software artefacts that satisfy the requirements (post requirements specification traceability), while the process-based traceability covers the whole development and maintenance process as well as the management process.

To fully exploit the potential knowledge in requirements traceability, it is essential that the structure of the information reflects the goals and methods of the development team. A convenient way to achieve this is to explicitly model the information according to product and process. In this way, a requirements management tool becomes process and product aware.

Awareness of the product model enables customised traceability and analysis, focusing on the knowledge relevant to certain project goals. Awareness of the process model adds further value, enabling to better investigate and understand the causal relations of underlying traces, (Lavazza and Valetto, 2000). Furthermore, process-based traceability provides added value, since it allows taking into account the process knowledge for the analysis of requirements change: it relates a change to the corresponding work through its impact.

The key issues of RE related to traceability that can affect the quality and effectiveness of RE in implementation are listed in Table 5.1.

Process models describe the activities that exist within a business process. A process model can be developed to a very fine or very rough degree of detail. In fragmented but multidisciplinary work environments such as the construction industry, process awareness is very low and process redesign is crucial for the implementation of CIE in such sectors. The following steps can be followed for the success of the re-engineering processes with IT (Betts 1999):

- Develop business vision, goals and process objectives.
- Identify the processes to be re-engineered.
- Understand and measure existing processes.
- Identify IT levers that will help push the changes.
- Design and build a prototype of the new processes.

Table 5.1 The key issues for requirements engineering that are related to traceability.

	Key issues
K1	Using traceability for process modelling
K2	Using traceability for product modelling
K3	Proper structuring and formalisation of the knowledge captured with traceability
K4	Customising traceability based on the organisation's and project's specific goals
K5	The ability to handle the inherent complexities in the technical, managerial and contractual dimensions in the RE process
K6	Conducting requirements management with process and product awareness
K7	Ability to measure the development process to support quantitative analysis
K8	Quantitative assessment of the impact of requirements changes and quantitative estimation of cost of the development activities

There has been some research about redesigning the construction activities as processes, such as the Process Protocol Map (Cooper and Wootton, 1998), based on which process models of the CIE systems should be defined. As mentioned earlier, business process redesign is another challenge for effective implementation of CIEs in the construction industry because the current work structure of the construction industry is not suitable for CIE implementation.

Process models are used in Industry Foundation Classes (IFC) specification development projects as the means to discover and capture the information content of a business process and how that information is to be exchanged between participants in the process. However, IFC specification developments are mainly dominated with product modelling. New innovative systems often result from a change in the basic process of the business, although sometimes the process can only be changed by the introduction of a new system. Whatever the case, the change should be process led and not IT led.

5.2.2 Goal-oriented requirements engineering

Goal-oriented RE is concerned with the use of goals for eliciting, elaborating, structuring, specifying, analysing, negotiating, documenting and modifying requirements. Goals can be formulated at different levels of abstraction, ranging from high-level strategic concerns to low-level technical concerns. Goals also cover different types of concerns: functional concerns and non-functional concerns, such as safety, security, accuracy, performance etc. (Sommerville and Sawyer, 1997;

Lamsweerde, 2001). There are reasons why goals are important in the RE process. Some are listed below.

- Achieving requirements fulfilment is a major RE concern. Goals provide a precise criterion for fulfilment of a requirements specification.
- Avoiding irrelevant requirements is another major RE concern. Goals provide a precise criterion for requirements pertinence.
- Explaining requirements to stakeholders is another important issue. Goals provide the rationale for requirements in a way similar to design goals in the design process. A requirement appears because of some underlying goal, which provides a base for it.
- Goal refinement provides a natural mechanism for structuring complex requirements documents for increased readability.
- Managing conflicts among multiple viewpoints is another major concern. Goals have been recognised to provide the roots for detecting conflicts among requirements and for resolving them eventually.
- Goals drive the identification of requirements to support them; they are shown to be basic driving forces for a systematic requirements elaboration process.

Some advantages of goal-oriented RE are as follows:

- Object models and requirements can be derived systematically from goals.
- Goals provide the rationale for requirements.
- A goal graph provides vertical traceability from high-level strategic concerns to low-level technical details; it allows evolving versions of the system under consideration to be integrated as alternatives into one single framework.
- Goal refinement structure provides a comprehensible structure for the requirements document.
- Alternative goal refinements allow alternative system proposals to be explored.
- Goal formalisation allows refinements to be proved correct and complete.

The key issues related to goal-oriented requirements are listed in Table 5.2.

The construction industry is a multidisciplinary sector with many different stakeholders involved in the operation/completion of a construction project. It is acknowledged that the current work structure is not effective and is profitable with many drawbacks, such as a

Table 5.2 The list of key issues related to goal-oriented requirements engineering.

	Key issues
K9	Completeness of the requirements specification
K10	Ability to keep requirements relevant and eliminate irrelevant requirements
K11	Enabling a shared understanding between users and developers
K12	Readability of the complex requirements engineering document
K13	Ability to resolve the inconsistencies in requirements
K14	Continuous traceability in the RE process

lack of communication and coordination, duplications, legal issues, distribution of the stakeholders, lack of process awareness etc. Overall, the CIE systems are aimed at improving the quality of construction and increasing the profit margins and work efficiency. In order to achieve this aim, the requirements of the construction stakeholders, who work on very complex, isolated platforms, at professional, business and project levels, must be properly captured.

In order to cover the requirements of all the stakeholders who work in a lean process, completeness and coherency of their specific requirements is important. Through a shared understanding amongst the stakeholders, possible conflicts may be avoided by keeping to pertinent requirements and discarding any irrelevant requirements. Because these stakeholders work in a complex environment, their requirements and expectations from the CIE systems will be complex, in which case, requirements specification will also be complex. Therefore, readability of the specification is crucial for shared understanding between the users and the developers.

Goals are good tools to capture the end users' requirements because construction stakeholders in general are not particularly familiar with the sophisticated IT solutions such as CIE. Therefore, because the goals are understood by everybody they can form a good platform to establish a shared understanding of what to do in CIE development. Subsequently, it helps to discard the inconsistencies in the requirements specification. It is inevitable that requirements will change, as the essential complexity grows in the development. This will be managed and simplified by means of continuous traceability in the RE process.

5.2.3 Essential and incidental complexity in requirements models

RE is an activity in understanding and problem solving. Contemporary literature describes the process as a systematically evolutionary development of requirements models during which requirements evolve and undergo frequent changes that often cause major difficulties during the development process. At the same time, the incremental evolution of the RE process results in the decrease of incompleteness, and thus the ambiguity of the problem is reduced (Loucopoulos and Karakostas, 1995). A deep understanding of the complexity of the requirements model and its dynamics is critical in improving the management of the RE process. There are two different types of complexities in requirements models. These are the essential and incidental complexities (Nguyen and Swatman, 2000).

The essential complexity denotes the inherent understanding of the problem space, whereas incidental complexity spans from the poor fit between the structures of the model and the structure of the world, which the model aims to represent. The evolution of the requirements model embraces both the growth of the essential complexity throughout the discovery of the problem space and its growth and the shrinkage of the incidental complexity as the model undergoes a large number of changes (Moody and Shanks, 1998; Roseman, 1998). The key issues related to complexity in RE are listed in Table 5.3.

The inherent understanding of distributed, collaborative integrated information systems for the multidisciplinary sectors such as the construction industry does exist between practitioners and academics. For example, in the Distributed Virtual Workspace for Enhancing Communication within the Construction Industry (DIVERCITY) project, the inherent understanding was very low but throughout the development it evolved through continuous RE activities. However, in general, at business, project and professional levels, the inherent understanding of the CIE concepts and possible benefits are still very immature and need to be improved through effective RE processes and subsequently successful CIE system demonstrations.

With regard to incidental complexity, the RE process should cover it because the CIE systems require business process redesign too and

Table 5.3 The key issues related to complexity in RE activities.

	Key issues
K15	The extent to which there is an inherent understanding of the problem space
K16	The extent of the fit between the structure of the requirements model and the structure of world
K17	Completeness and unambiguousness of RE model

the current actual work practices are not suitable and include high complexity, which is not favourable for the effective implementation of CIE. In Chapter 4 this issue is justified through uncontrollable parameters existing in the construction industry, whereas RE is classified as controllable parameters to contribute to the implementation of CIE in the construction industry.

Since the construction industry is a complicated one, this complexity will be reflected in the requirements specification. To make matters worse, it is inevitable that there will be conflicts between the needs of different stakeholders, which will add an extra layer to the complexity of the requirements specification. Consequently, the completeness and unambiguousness of the requirements specification and model is critical.

5.2.4 The measurability of quality requirements

The root-cause analysis conducted by researchers, for example by Jacobs (1999), aimed to define actions to increase the quality in terms of fault density and to decrease the lead-time in the RE processes. This analysis identified that 'No common understanding of what to do' was the main obstacle for decreasing fault density and lead-time (Jacobs, 1999). A number of shortcomings in producing requirements specifications were also identified, which are as follows:

- Requirements specifications were long and monolithic, based on narrative text. The requirements were unstructured. There was no conceptual difference between functional and non-functional requirements. There was no hierarchical structure; all requirements were just put into a long list.
- Quality requirements were not precisely defined. Typical examples were 'increase the usability' or 'the system has to become more reliable'.
- Costs in terms of elapsed time, man-hours or money were not stated as a requirement at all.
- Requirements specifications were not maintainable. They include inconsistencies, which were not detected. Introducing new requirements and analysing their relations to other requirements was not possible.
- Only parts of the requirements were tested. Non-functional requirements were especially hardly checked.
- As there are no rules on how to write requirements specification, inspections could not check these rules.

- It was not possible to trace the requirements, neither to their source nor to their target.

In order to overcome the above problems in requirements modelling, Jacobs (1999) proposed the Gilb Style, which is as follows:

1. **Structuring information:** To implement these concepts, a certain level of structure or formalism is needed. In Gilb's method, requirements are divided into the following:
   - Functional requirements
   - Quality requirements
   - Constraints
   - Cost requirements.

   Functional requirements describe what a system does. Quality requirements describe how well the defined functionality has to work. Quality requirements specify the soft and analogue element of functional requirements. To be precise, quality requirements have to be quantified. Typically, they are divided into several sub-requirements to cover different aspects of global terms. For example, the quality requirement *usability* could be divided into the requirements *easy to use, easy to learn, fast to use*, etc.

   Constraints form a restriction that must be taken into consideration when designing the complete product. Constraints restrict the solution space for the requirements. Cost requirements are a special case of quality requirements. The term *cost* covers not only money but also all kinds of resources like time, space, people, satisfaction, reputation, etc. As these resources are always limited, they are a crucial part of the requirements specification.
2. **Quantification:** It is necessary to make requirements measurable. It is more difficult but essential to make quality requirements measurable. *Gist, scale, meter, past, record, must, plan and wish* are the concepts that were highlighted by Jacobs (1999) in the development of requirements specifications.

   Gist is a rough summary of the requirements. It is useful to have an overview of all requirements without going into detail. *Scale* defines the unit in which the requirements have to be measured. Scale defines a set of measures to express notions of qualities and costs. *Meter* defines the way in which the measurement will be performed. It can, for example, specify a test. *Past* and *Record* are benchmarks. *Past* is a value that is typical for products developed in the past. The end user compares new products against this already achieved level. On the other hand, *record* represents the best value that has been achieved for this meter. To target a market or product on the market it is crucial to know both values.

In contrast to *past* and *record*, which are historical values, *must*, *plan* and *wish* envisage the future. They characterise the system that is to be built. As the names indicate, *must* represents the minimal goal level. Failure to reach the *must* level indicates a failure level for the system. *Plan* is the envisaged level for a requirement. However, not reaching the *plan* level indicates a failure only for this requirement and not for the whole system. Finally, the *wish* level represents the level that is aimed for under-optimal circumstances. The *wish* value is usually not committed to because of resource limitations.

3. **Sourcing requirements:** To trace the requirements to their source, it is essential to know where the requirements are coming from.
4. **Tagging requirements:** Each requirement, including constraints and costs, and each assumption has to be uniquely identifiable. To assure this, unique tags have to be assigned to each requirement.

The key issues related to the measurability of quality requirements are listed in Table 5.4

As mentioned earlier, a shared understanding between the construction organisations at the organisational level, a shared and compromised understanding between the construction stakeholders at project level, and a shared understanding of what they exactly need from a CIE system is crucial for the successful development of CIEs. For example, the DIVERCITY requirements capture a process that is evolved on the basis of the evolving shared understanding.

Apart from functional requirements, providing a user-friendly system will make the stakeholders willing to employ such a system in their organisation. CIE technology provides sophisticated functionalities,

Table 5.4 The key issues related to measurability of quality requirements.

	Key issues
K18	The extent of common and shared understanding of what to do
K19	The extent of structuring the requirements specification for a clear picture of user requirements
K20	Conceptual differentiation of functional and quality requirements
K21	The level of impact of cultural attitude on the requirements engineering process
K22	The extent of quantification of non-functional requirements
K23	The extent of traceability in requirements for sourcing the information
K24	The extent of thoroughness given to the constraints and cost requirements

but the ease of use of these functionalities, the user-friendly interface and the other quality aspects should be properly taken into account as they are important in convincing the decision makers to adopt the CIE systems in their workplace.

There used to be a negative attitude in the early culture of employing RE in the earlier CIE system development, but its impact is becoming lower, whilst the awareness of the stakeholders about the construction CIE system development is rising. Avoiding these attitudes is also important for innovative, distributed, collaborative system development such as the CIE systems.

5.2.5 The requirement fundamentals

The requirements specification and updating and verifying the requirements through continuous testing with users' tests are amongst the fundamentals of RE. Robertson (2000a) raised the quality gateway that defines test points that a requirement has to pass in order to meet the requirements specification.

The set of requirements tests cover relevance, coherency, traceability, completeness, consistency, connectivity and other qualities that successful requirements must have. The starting point of the tests is that each requirement must have a number of components, one of which is a fit criterion.

Fit criteria are used to test whether it is possible to classify any given solution into one that satisfies or does not satisfy the requirement. The aim is to trap requirements-related defects as early as they are identified and to prevent incorrect requirements being incorporated in the design and implementation where it will be more difficult and expensive to find and correct.

There is a critical problem about the requirements specification, which is a vague term in current use. Some specifications contain purely functional requirements, others include non-functional requirements and some include business goals, whilst some concentrate on the solution. However, without consistent agreement on the intended specification, it is impossible to test the quality of its contents. In order to build the product and deal with changes, it is required to keep track of connections – for example, which requirements are implemented by which product use case and which product use cases are necessary because of which business events. To test the completeness and traceability of the requirements, it must be possible to identify these connections.

The aim of the quality gateway is to discover problems such as unambiguous requirements, missing requirements and inconsistent requirements. If the requirements have been defined in a consistent and testable way, then people with testing expertise can put the requirements through the quality gateway tests. The following are some test points to build into quality gateway.

- Does the specification contain all the requirements components?
- Does the specification contain all the relationships between components?
- Is each requirement uniquely identifiable?
- Is each goal measurable?
- Does the event list agree with the work context?
- Are all the event inputs and outputs defined?
- Can you quantify why each constraint exists?
- Is each constraint measurable?
- Are the user characteristics defined?
- Are the stakeholders' responsibilities defined?
- Can you trace the product use cases back to the business events?
- Are the terms used in the fit criteria defined in the requirements specification?
- Is it possible to write a cost-effective test to prove or disprove any solution to this requirement according to its fit criterion?
- Are conflicts between requirements identified?
- Is each requirement really a requirement or is it a solution?

If a project team is aware of the test points, then this awareness will help requirements engineers to improve the quality of the requirements. The quality gateway test is a set of points in the project where testing expertise is used to see how well the development is progressing before engineers issue a document and allow errors to pollute the product.

The key issues related to requirements fundamentals are listed in Table 5.5.

Involvement of the stakeholders in a CIE development is crucial because they will drive the development and highlight the key issues of the inter-disciplinary nature of their work in such construction projects. Testing the requirements with the stakeholders through quality gateways can be a good method of verifying and validating the requirements throughout the development. This is because it will enable the stakeholders to gain an improved understanding of CIE and provide the relevant and correct requirements and leave out the irrelevant requirements.

While testing the early prototypes in the development life cycle, it is also very efficient to update and validate the requirements as well as the system itself. Furthermore, testing should also cover the

Table 5.5 The key issues related to requirements fundamentals.

	Key issues
K25	Testing the requirements for consistent and accurate specification
K26	The extent of relevance, coherency, traceability, completeness, consistency and connectivity in the requirements model
K27	The level of balance between functional requirements and the other requirements in the requirements specification
K28	The quality of structure and documentation of the requirements specification for better readability
K29	The extent of capability to trap requirements-related defects as early as they are identified and to prevent incorrect requirements from being incorporated in the design and implementation
K30	Involvement and commitments of the stakeholders

quality requirements to make the system user friendly and comply with the quality requirements. Furthermore, testing can help to simplify the incidental and essential complexities in the requirements models, which also bring about a shared understanding between the stakeholders and developers. Since there will be different parties from different backgrounds in the user groups of the development team, requirements specification should have the same readability for everybody. To do so, the same template and language must be used in the requirements specification and each time the specification is updated, versioning should be done for change management.

5.2.6 Identifying and involving the stakeholders

The major factor that affects the success of a project is having the right people involved in the right activity at the right time. It is acknowledged that the need to identify the project's stakeholders and to find ways of appropriately involving them throughout the life of the project is crucial (Robertson, 2000b). The first step towards discovering all the requirements is to understand who the stakeholders are, what roles they are expected to play, how much involvement they need to have and how committed they are to the project. In short, it is required to understand the project's sociology.

Traditionally, concerns about the stakeholders have focused on the developers of the product and the users of the product. Robertson (2000b) advised that it is required to look more widely at the people whose knowledge contributes to the project's success instead of talking to end users and developers and identified a number of different types of stakeholders whose participation is potentially necessary for a project's success. For example, stakeholders are clustered into two: producers and supplementary stakeholders.

Producers are the people who are assigned to work on the project and deliver some product or a mixture of products that satisfy the requirements. Producers are typically people such as project managers, developers, business analysts, key business people, marketing specialists, research and development specialists, designers, testers, and programmers. Producers should be people who form a cohesive team collaborating to produce a shared product. They should be in close contact with each other.

The supplementary stakeholders are not assigned to work full-time on the project. They have the knowledge that is required by the producers. Each of these different types of stakeholders can provide the producers with feedback about their own particular knowledge and concerns. However, in order to stimulate that feedback, the producers need to provide a feedback mechanism that is appropriate for the particular stakeholder, which means understanding the individual stakeholders and understanding what knowledge the project needs from them, when they need to be involved, how they relate to each other and how their involvement will affect the success of the project.

Key issues related to identification and involvement of stakeholders are listed in Table 5.6.

Since the concept of CIE is for the multidisciplinary work environments, it is critically important to involve the right stakeholders who can provide vital contribution to the development and implementation of the CIE systems. For example, decision makers for implementation should be involved in the development. Subsequently, they will be convinced and willing to utilise the CIE toolkits in their workplace and furthermore, they will urge the other people and organisations to employ the same technology in their organisations. For example, on a professional and project level, the right construction practitioners should be involved in the development at the right time in order to address the key bottlenecks they have experienced in their own professions, and in a construction project it is their duty to do so.

Table 5.6 Key issues related to identifying and involving the stakeholders.

	Key issues
K31	Involvement of the right stakeholders at the right time in the right subjects
K32	A feedback mechanism through which there is a two-way communication between the producers and stakeholders in order to stimulate feedback

5.2.7 Reconciling software requirements and architectures

Since the software systems of today are characterised by increasing size, complexity, distribution, heterogeneity and lifespan, at an early stage they demand careful capture and modelling of requirements (Nuseibeh and Easterbrook, 2000) and architectural designs (Shaw and Garlan, 1996) before the low-level system details begin to dominate the engineers' attention, and significant resources are expanded for system construction. Evolving and elaborating system requirements into viable software architecture to satisfy those requirements is a difficult task. Engineers face some critical challenges when trying to reconcile requirements and architectures (Grunbacher *et al.*, 2001).

- Requirements are frequently captured informally in a natural language. On the other hand, entities in a software architecture specification are usually specified in a formal manner.
- System properties described in non-functional requirements are commonly hard to specify in an architectural model.
- Certain requirements can only be understood after modelling and even partially implementing the system architecture (Nuseibeh, 2001).
- Mapping requirements into architecture and maintaining the consistency and traceability between the two is complicated since a single architecture element may have numerous non-trivial relations to various requirements.
- Real world, large-scale systems have to satisfy hundreds of requirements. It is difficult to identify and refine the architecturally relevant information contained in the requirements because of the huge scale.
- Requirements and the software architecture emerge in a process involving heterogeneous stakeholders with conflicting goals, expectations, and terminology. Therefore, supporting the different stakeholders' demands is critical to find the right balance across these divergent interests. (Grunbacher and Briggs, 2001).

Requirements describe the aspects of the problem to be solved and constraints on the solution. Requirements are derived from the concepts and relationships in the problem domain. They reflect the interests of system's stakeholders. Requirements deal with concepts such as goals, options, agreements, issue, conditions and desired system features and properties. Although requirements are easier to understand by people, they frequently lead to ambiguity, incompleteness and inconsistencies. On the other hand, architectures model a solution to the problem described in the requirements. Software architectures provide high-level abstractions for representing the structure, behaviour and key properties of a software system.

Architecture deals with components, which are computational, and data elements in the software system (Perry and Wolf, 1992). Components and connectors are composed into specific software system topologies. Finally, architectures both capture and reflect the key desired properties of the system under construction (e.g. reliability, performance, and cost). An approach called the *Component-Bus-System-Property* (CBSP) was introduced by Grunbacher *et al.* (2001) and helps to refine a set of requirements into architecture by the classification of architectural dimensions. In this iterative process, CBSP provides requirements to an architecture model connector that helps to concurrently evolve requirements and architecture models (Nuseibeh, 2001).

The key issues related to reconciling requirements and architecture are listed in Table 5.7.

This section has addressed the transition from the requirements specification to the system architecture. Depending on the success in that transition, the system will reflect the requirements of the stakeholders. Because there are heterogeneous stakeholders involved in a CIE development, it is a big challenge to capture and model these requirements consistently and coherently while minimising the conflicts. In addition to the achievement in modelling these complex requirements, transforming the complex requirements model into system architecture is very critical in the CIE development. The transformed system architecture should comply with the process model defined in the requirements model and the product model.

5.2.8 Barriers to uptake of requirements engineering

It became obvious that although there have been numerous research and development (R&D) projects at both the national and European

Table 5.7 The key issues related to reconciling requirements and architecture.

	Key issues
K33	The extent to bridge different levels of formality
K34	Modelling non-functional requirements
K35	Maintaining evolutionary consistency
K36	The fit between the system architecture and the system requirements and the way the user works
K37	Handling scale and complexity
K38	Involving heterogeneous stakeholders

levels, industrial uptake from R&D projects has rarely lived up to expectations. To understand the difficulties with industrial uptake, there are two critical goals to be considered (Morris *et al.*, 1998):

- To identify and prioritise the problems encountered on RE R&D projects with industrial uptake
- To identify and prioritise recommendations for improving industrial uptake.

The problems identified from research studies are clustered into four groups. These are training, inherent complexity, internal business integration and business culture. These aspects are briefly explained below (Morris *et al.*, 1998; Arayici, 2004).

Training: Three types of difficulties are identified. First, the training costs for introducing new methods are very high. There is always pressure on resources, which limits the opportunities to train and retrain. Second, as companies have to work with their current skill base and within their source limits, they are compelled by market pressures to make effective use of them. This, in the short term, makes re-training difficult. Finally, pressures to standardise methods create resistance to new approaches.

Inherent Complexity: Although RE is a central part of the development process, it is the area in which most problems arise and is the area where problems are the cheapest to fix. However, there are still technical problems. There were a number of explanations for the difficulties in RE:

- The various stakeholders providing requirements and the number of requirements make integration into a coherent specification a complex and difficult process.
- There can be conflicts in the requirements arising from different viewpoints and there is the potential for ambiguity and failure to communicate.
- As complexity tends to increase disproportionately, methods must be scalable and able to handle large sets of requirements.
- Requirements are generated, not just through the requirements elicitation process, but also during requirements analysis and the downstream of the RE process.
- There is no evidence that production of engineering diagrams, drawings and formal methods form an appropriate way to specify systems.

Internal Business Integration: The third problem area is concerned with business integration. Difficulties arise with integrating the RE processes into particular organisations because of business quality standards. There are three reasons for these difficulties, which are as follows:

- The position of RE within the whole development process means that significant changes in the way requirements are approached will affect the rest of the development. New methods imply that all stakeholders have to change their way of thinking and working. Although there is pressure to change the more traditional development process, the context of the system design and development process creates resistance against changing methods and techniques.
- There is no guarantee that the adoption of novel techniques will continue to preserve or improve product quality, usability and robustness. Measures of the impact of the methodology, the tools and the improved specification on the whole project are difficult to obtain.
- There is a management culture of distributing responsibility. Adoption of RE methods, with extensive tracing relationships, will result in the assignment of responsibility of particular requirements to particular stakeholders. This is incompatible with the current practice of distributed responsibility.

Business Culture: In general, management support for RE is low. There is a lack of familiarity with the RE techniques as an independent well-established and approved discipline. Both user and stakeholders have low maturity in this area. There are three reasons identified for this problem.

- Firstly, management does not give full attention to RE because the cost of implementation and execution is too high. At the start of a project, there is usually a shortage of financial resources. It is therefore difficult to expend considerable resources on the RE.
- Secondly, there is a need to show that the production costs can be kept within bounds. Formal methods in particular would be expensive to use.
- And thirdly, not all development processes are the same. Different kinds of solutions are therefore required.

Key issues related to the uptake of RE are listed in Table 5.8.

Barriers to the implementation of RE techniques exist in the software industry, which is a similar condition for the implementation of CIEs in the construction sector. Therefore, the RE process for the CIE development should be comprehensible because it can be an enabler for the implementation of CIEs in the construction industry. One of the most difficult problems with improving the use of IT in construction

Table 5.8 Key issues related to barriers against the uptake of RE.

	Key issues
K39	The extent of coverage of cost and benefits justifications
K40	Scalability of RE method to handle large sets of requirements
K41	The clarity and coherency of the requirements specification derived through requirements elicitation from many different stakeholders
K42	Continuity of requirements capturing through the requirements elicitation, requirements analysis and downstream of the RE process
K43	The familiarity of the end users and stakeholders with the RE techniques
K44	The extent of cost implementation and execution
K45	The equivalence between the available financial resources and the required financial resources
K46	The maturity of the end users and stakeholders

is overcoming the implementation problem, which requires a strategy definition. Factors that need to be considered are as follows (Betts, 1999):

- Business processes are relevant to process modelling in RE in the CIE development. As Betts (1999) emphasised, new innovative systems should be process led and not IT or product led.
- Recruiting the project teams is relevant to defining the producers and the other stakeholders who will contribute to the development. Team members need to have the ability to convey progress to and seek the advice and approval of their peers. However, whilst it is not always possible to represent or obtain the views of every interested party, the chosen team members should fairly reflect the views of all interested parties.
- System integration implies an integrated computer environment that consists in implementing more than one solution. Project data passes from one solution to another through an integration platform. Exchange of information and data is conducted by means of IFC.
- Timescale and resources are important because once the implementation process starts, the project should be completed as soon as possible, within the allocated timescale and resources.

5.3 Measuring the Success of Requirements Engineering Process

A prerequisite to improving RE practices is a fundamental understanding of RE processes and of the factors that lead to their success or cause their failure. By identifying key factors through research studies by which RE techniques can be assessed the success of RE techniques can be observed. Research studies by Emam and Madhavji (1995) and Arayici *et al.* (2006) compiled 34 key factors, which cover the two most important categories of RE success. These two categories are the quality of RE products and the quality of RE process.

As a result of a large number of interviews with 30 experts, each of whom was involved at least once in the RE phase in different roles, the outcome of the first step was a set of 34 criteria for the assessment of RE success. Subsequently, dimensions of RE success were identified and the criteria were prioritised by Emam and Madhavji (1995).

The cluster analysis of the criteria is concluded with five clusters as shown in Table 5.9. These interpretations were derived from the information provided by the interviewees. Since two of the clusters directly concerned how good the outputs of the RE processes were (quality of architecture and quality of cost–benefits analysis), they were grouped into one general cluster, which is the quality of RE products. Furthermore, since two other clusters were related to the quality of the RE service dimension (user satisfaction and commitment and match of the CIE system with companies), they were also grouped together.

The resultant three categories of RE success are as follows:

1. **The cost effectiveness of the RE process:** This category assesses whether a reasonable amount of resources were consumed during the RE phase.
2. **Quality of RE products:** The quality of RE products covers the quality of the major documents that are produced during the RE phase – quality of architecture and quality of cost–benefits analysis.
3. **Quality of RE process:** This category concerns the service provided by the RE staff to the users. The quality of this service is reflected in the extent of user satisfaction and commitment and match of the CIE solution with the construction companies. It is critical to provide a service that satisfies the users and to properly manage their expectations.

A total of 34 criteria for assessing the success of the RE process have been identified and listed according to their priority. Categories are also ordered in their priority (Emam and Madhavji, (1995); Arayici, 2004).

5.4 Comparative Analysis and Evaluation

This section summarises and also represents the comparative analysis of the approaches in this chapter to extract the set of evaluation

Table 5.9 Criteria for the evaluation of requirements engineering practices (adapted from Emam and Madhavji, 1995).

Category: Quality of Requirements Engineering Process (Priority: 1)

Theme 1: Match of the CIE system with construction companies

C1	The capability of companies to make changes for uptake
C2	The match between CIE system and strategic orientation of the companies
C3	Senior management support for changes for uptake
C4	The willingness of companies to make changes for uptake
C5	The match between CIE system architecture and corporate structure of organisations in the construction projects
C6	The match between CIE system and technical orientation of the companies

Theme 2: User satisfaction and commitment

C7	Level of understanding of end users about what the CIE system do and will not do
C8	The willingness of users to defend CIE system in front of executive management
C9	User consensus on CIE system
C10	Awareness of users about the business changes for uptake
C11	The relationship between stakeholders and the RE team
C12	The reaction of users to the cost estimation
C13	The fit between the system structure and the way the construction stakeholders practice
C14	Whether users approved all the RE documentation
C15	The level of match between functionality of alpha release and user expectations

Category: Quality of RE Products in CIE System Development (Priority: 2)

Theme 3: Quality of cost–benefits analysis

C16	Completeness of the cost–benefits analysis
C17	The level of convincing of executive management that expected benefits can be materialised
C18	Fit between the available funding and the funding required for uptake
C19	The amount of benefits to the construction stakeholders by uptake
C20	The accuracy of the cost estimates relative to the accuracy required by organisations
C21	To what extent the cost–benefits analysis follows the accounting procedures of organisations
C22	Appropriateness of approaches taken to quantify intangible benefits when the system is implemented

Theme 4: Quality of the CIE system architecture

C23	Clarity of the links between the process models and product models and the system objectives

(continued)

Table 5.9 (*Continued*)

C24	Clarity of the business process in the system structure
C25	Clarity of links between the process models, data models and the key issues
C26	Attention given to the alternative solutions for the CIE system to be developed
C27	Clarity of links between weaknesses and strengths of information and communication technology (ICT) systems in use and weaknesses and strengths of the CIE system to be developed
C28	Adequacy of diagnosis for ICT solutions in use
C29	The level of extent to key issues have been resolved
C30	The level of compliance of process and data models to the rules of modelling
Category: Cost Effectiveness of RE Process in CIE Development (Priority: 3)	
C31	Cost and effort on the RE process
C32	Proportion of the cost of RE compared to the estimated total system development cost
C33	Amount of changes in the RE documentation
C34	Amount of deliverables that were not used in system designing of CIE

criteria, which will form the basis of evaluation of RE in the case study projects explained in Chapter 3.

Research studies whose objectives were to identify the key criteria that will affect success or failure of the RE activities have been represented in the previous section. On the other hand, some other researchers and engineers have used approaches from various points of view to improve RE, which have also been summarised in Section 5.2. Key issues that have influences upon RE have been extracted from these approaches. Furthermore, the suitability and relevance of these key issues have been discussed under each approach in the same section.

In this section, the criteria from Section 5.3 and key issues from Section 5.2 are analysed and comparatively matched. This comparative analysis and evaluation confirms the validity of the set of criteria from Section 5.3 for the evaluation of the case studies explained in Chapter 3.

In Table 5.10 the key issues and criteria are matched to each other. Accordingly, the set of criteria is filtered from 34 criteria.

The comparison in the above table is done in a cause and effect manner. That is to say, key issues identified are to improve the RE. On the other hand, criteria determined in Section 5.3 represent the end results or effects after adapting some particular key aspects in an RE approach that has already been experienced. Therefore, while the

Table 5.10 The benchmark between the criteria and the key issues to confirm validity of the set of criteria from Section 5.3.

	C1	C2	C3	C4	C5	C6	C7	C8	C9	C10	C11	C12	C13	C14	C15	C16	C17	C18	C19	C20	C21	C22	C23	C24	C25	C26	C27	C28	C29	C30	C31	C32	C33	C34
K1					*																		*	*	*									
K2					*																		*		*									
K3																							*	*										
K4		*				*																	*											
K5																								*	*									
K6							*																*		*									
K7																*															*			
K8																				*					*							*	*	
K9														*															*					
K10							*				*																							
K11									*		*																							
K12							*																											
K13											*																		*					
K14																							*	*	*									
K15									*																*									

(continued)

Table 5.10 (*Continued*)

	C1	C2	C3	C4	C5	C6	C7	C8	C9	C10	C11	C12	C13	C14	C15	C16	C17	C18	C19	C20	C21	C22	C23	C24	C25	C26	C27	C28	C29	C30	C31	C32	C33	C34
K16						*							*										*	*										
K17														*															*					
K18							*		*					*																				
K19							*																											
K20																													*					
K21																														*				
K22																						*												
K23																									*				*					
K24												*																	*					
K25																						*			*									
K26													*																					
K27																						*												
K28														*																				
K29																						*												
K30								*																					*					
K31								*																					*					

K32											*																		*					
K33					*																									*				
K34																						*							*					
K35																													*	*				
K36		*			*								*											*										
K37																								*						*				
K38								*																					*					
K39																*																		
K40																								*						*				
K41																							*	*										
K42																							*		*				*					
K43	*		*					*	*	*																								
K44												*																			*			
K45																			*		*													
K46	*		*							*																								

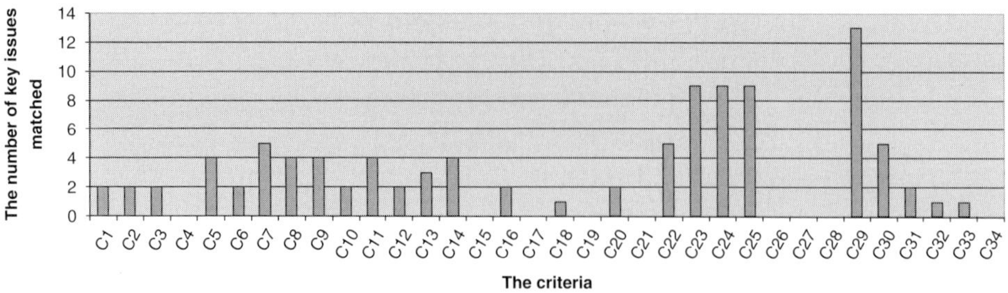

Figure 5.1 The level of fit in comparison between the criteria and key issues.

comparison is being done, the possible causes of each criterion in the criteria list have been sought in the list of key issues.

The graph in Figure 5.1 represents the level of fitness between the 34 criteria and the 45 key issues. Although the number of key issues is more than the number of criteria, all the key issues cannot compromise all the criteria. This is attributable to the fact that the research studies for criteria identification represented in Section 5.3 in order for measuring the success of the RE techniques have more comprehensive results than the key issues identified from the other research studies on improving the RE processes in this chapter.

While the key issues strongly addressed the technical aspects of RE process in the development life cycle at the project level only, they weakly address the managerial, organisational, cost–benefit and implementation aspects at the professional and organisational levels. The criteria address the managerial, organisational, cost–benefits and implementation perspectives as well as the technical aspects. In other words, the set of criteria addresses the issues for the success of the RE process in organisational, professional, business and project levels according to the determined priority. Figure 5.1 illustrates the summary of comparison between the criteria and key issues.

Criteria from C1 to C6 reflect the success of the RE technique at the organisational level, which is inadequately addressed in the key issues, as shown in the graph in Figure 5.1. On the other hand, criteria from C7 to C15 denote the user satisfaction and commitment for implementation, which is addressed consistently in the key issues. Therefore there is a persistent fit between the criteria and key issues in that respect. However, in the key issues, quality of the cost and benefit analysis has been disproportionately addressed, which comprises the criteria from C16 to C22. Unlike the overall fit of the criteria with the key issues in regard to the quality of cost–benefit analysis, C22 is matched fine with the five key issues in respect to the quality of cost–benefit analysis.

In regard to the system architecture, criteria from C23 to C30 are clustered in this group in the quality of architecture. It is realised from the graph that the key issues largely address the quality of the architecture at project level. Especially, the number of the matched key issues with C23, C24, C25, C29 and C30 are relatively higher than the others. However, there is no match for the criteria C26, C27 and C28 although these criteria are grouped in the dimension of the quality of architecture. This is because these three criteria entail the benchmarking with alternative solutions or existing systems in use or the other commercial off-the-shelf packages. However, the benchmarking has not been highlighted in the determination of the key issues.

The criteria from C31 to C34 are classified as cost effectiveness of the RE processes. Although there is a match between the key issues and criteria apart from C34, it is relatively poor compared with the other matches in the other clusters, which is attributable to the fact that nearly all criteria in this category require benchmarking. However, benchmarking has not been addressed in the key issues. For example, the same key issue matches both C32 and C33. The dimension of cost effectiveness is ranked three; the thoroughness and emphasis given that dimension in the key issues have also been less than the other dimensions.

As a result, the set of criteria encompass the whole key issues defined, which are related to CIE systems in discussion parts in each section. Subsequently, these criteria are sufficient to evaluate the requirements capture processes in the case study projects at organisational, professional and project levels. On the other hand, the key issues extracted from other research studies will be used for the assessment of the RE approach to be proposed for the development of CIE systems.

References

Arayici, Y., (2004), "Requirements Engineering in Innovative Systems Development for the Construction Industry", PhD thesis, SCPM, University of Salford, United Kingdom.

Arayici, Y., Ahmed, V., Aouad, G., (2006), "A Requirements Engineering Framework for Integrated Systems Development for the Construction Industry", Electronic Journal of Information Technology in Construction (ITcon), 11:35–55, www.itcon.org/2006/3.

Basili, V., Rombach, H.D., (1988), "The TAME Project: Towards Improvement-oriented Software Environments", IEEE Transactions on Software Engineering, 14:758–773.

Betts, M., (1999), "Strategic Management of IT", London, Blackwell Science, ISBN 0 632 04026 2.

Brooks, F.P., (1987), "Essence and Accidents of Software Engineering", IEEE Computer, 20:10–19.

Carr, J., (2000), "Requirements Engineering and Management: The Key to Designing Quality Complex Systems", The TQM Magazine, 12(6):400–407, ISSN 0954-478X.

Cooper, K., Ito, M., (2000), "Experimental Evaluation of the Stimulus Response Requirements Specification Notation", Staffordshire, United Kingdom, EASE.

Cooper, R., Wootton, A., (1998), "Requirements Capture: Theory and Practice", Technovation, 18(8/9):487–511.

Davis, F., (1989), "Perceived Usefulness, Perceived Ease of Use, and User Acceptance of Information Technology", MIS Quarterly, 13(3):319–340.

Domges, R., Pohl, K., (1998), "Adapting Traceability Environments to Project Specific Needs", Communications of the ACM, 41(12):54–62.

Emam, K., Madhavji, N., (1995), "Measuring the Success of Requirements Engineering Processes", Second IEEE International Symposium on Requirements Engineering, York, United Kingdom, IEEE Computer Society Press, pp. 204–214.

Grunbacher, P., Briggs, B., (2001), "Surfacing Tacit Knowledge in Requirements Negotiation: Experiences Using EasyWinWin", Proceedings Hawaii International Conference on System Sciences, Hawaii, IEEE Computer Society.

Grunbacher, P., Egyed, A., Medvidovic, N., (2001), "Reconciling Software Requirements and Architectures: The CBSP Approach", Fifth International Symposium on Requirements Engineering, Los Alamitos, California, IEEE Computer Society Press, pp. 202–211.

Jacobs, S., (1999), "Introducing Measurable Quality Requirements: A Case Study", IEEE Joint International Conference on Requirements Engineering, Limerick, Ireland.

Keil, M., Cule, P.E., Lyytinen, K., Schmidt, R.C., (1998), "A Framework for Identifying the Software Projects Risks", Communication of ACM, 41:76–83.

Lamsweerde, A., (2001), "Goal Oriented Requirements Engineering: A Guided Tour", Invited Mini Tutorial Paper 5th IEEE International Symposium on Requirements Engineering, Toronto, Canada.

Lavazza, L., Valetto, G., (2000), "Enhancing Requirements and Change Management through Process Modelling and Measurement", IEEE Joint International Requirements Engineering Conference, Schaumburg, Illinois.

Loucopoulos, P., Karakostas, V., (1995), "System Requirements Engineering", Berkshire, United Kingdom, McGraw-Hill Book Company Europe.

Moody, D.L., Shanks, G.G., (1998), "What Makes A Good Data Model? A Framework for Evaluating and Improving the Quality of Entity Relationship Models", Australian Computer Journal, 30(3):97–110.

Morris, P., Masera, M., Wilikens, M., (1998), "Requirements Engineering and Industrial Uptake", IEEE Joint International Requirements Engineering Conference, Colorado Springs, Colarado.

Nguyen, L., Swatman, P.A., (2000), "Essential and Incidental Complexity in Requirements Models", IEEE Joint International Requirements Engineering Conference.

Nikula, U., Sajaniemi, J., (2002), "A Lightweight Combination of Proven RE Techniques", IEEE Joint International Requirements Engineering Conference, Essen, Germany.

Nuseibeh, B., (2001), "Weaving Together Requirements and Architecture", IEEE Computer, 34(3):115–117.

Nuseibeh, B., Easterbrook, S., (2000), "Requirements and Architecture: A Roadmap in the Future Software Engineering", 22nd International Conference on Software Engineering, ACM-IEEE, Limerick, Ireland, (Special Issue), pp. 37–46.

Perry, D.E., Wolf, A.L., (1992), "Foundations for the Study of Software Architectures", Software Engineering Notes, 17(4):40–52.

Porter, A., Votta, L., (1998), "Comparing Detection Methods for Software Requirements Inspections: A Replication using Professional Subjects", Journal of Software Engineering, 3(4):355–379.

Robertson, S., (2000a), "Requirements Fundamentals: The Basis for Effective Testing", The Atlantic Systems Guild Ltd, www.systemsguild.com.

Robertson, S., (2000b), "Project Sociology: Identifying and Involving the Stakeholders", The Atlantic Systems Guild Ltd, www.systemsguild.com.

Roseman, M., (1998), "Managing the complexity of multi perspective information models using the guidelines of modelling", Proceedings of the Third Australian Requirements Engineering, Geelong, Australia.

Shaw, M., Garlan, D., (1996), "Software Architecture: Perspectives on an Emerging Discipline", New Jersey, Prentice Hall.

Sommerville, I., Sawyer, P., (1997), "Requirements Engineering: A Good Practice Guide", Chichester, England, John Wiley & Sons.

Wu, M., (1990), "Selecting the Right Software Application Package", Journal of Systems Management, 41(9):28–32.

6

Requirements Engineering Approach in the Case Study Projects

6.1 Introduction

The construction industry is a large multidisciplinary industry, with multifaceted perspectives and requirements. It is extremely important to bridge the gap between end-user requirements and actual design and implementation of the underlying operational models of the distributed virtual workspaces. This is certainly true as the design relates to a new type of design artefact that will highly influence the traditional working methods and integration of design resources.

The broad industrial requirements need to be defined and expanded in more detail to capture the briefing and design requirements. It has a well-formulised requirements engineering (RE) approach amongst the aforementioned computer integrated environments (CIE) development projects. This chapter will subsequently elaborate and explain further the RE approaches in the case study projects, in particular Case Studies 3 and 4 that are explained in Chapter 3, because they have clear evidence of RE techniques and strategies as compared to Case Studies 1 and 2, which are more technology oriented.

In Case Studies 3 and 4, some techniques and tools were used to capture and model user requirements, such as Rational Rose, contextual design (CD) technique, storyboarding, etc. Furthermore, the understanding of requirements capture evolved in the project and turned into the requirements engineering process. After committing to code, RE was continuously carried out through incremental prototyping with the end-user tests. During the incremental prototyping, not only user requirements were improved, but the CIE system was also improved, according to the evolving user requirements specification.

This chapter explains RE approaches used in Case Studies 3 and 4 to capture the industry's requirement. In particular, the following are addressed:

- Evolving a shared understanding of the process aspect in the requirements efforts
- Continuous improvement
- Outputs from RE
- Reflections and critical analysis of the RE approach

6.2 The Need for the CIE System As a BIM Tool

In the past decade, construction companies have spent a great deal of effort and resources in improving their business processes. New forms of innovative project management, supported by recent IT developments, have appeared in response to ever-growing pressure from owners to complete projects on time and deliver high-quality buildings.

Although many sectors such as automotive, manufacturing and the service sectors have benefited very much from IT by improved competitiveness, the construction industry has had some difficulties, which have resulted in the industry lagging behind other sectors. The constraints that the construction industry has had may be summarised as follows:

- **The construction industry is fragmented by nature:** A high level of complexity in the workflow resulting from a large number of companies working on the same project increases inefficiency because of the fragmentation. For example, repeated processes, functions or duplications, due to the lack of communication and standardisation, result in waste and increase the lead time in the project's life cycle, as well as the extra cost to the client for non-value-added activities.
- **Design processes are separated from the construction processes:** This increases the uncertainty and incompatibility between the design solution and the actual construction, i.e. poor buildability. Increased uncertainty caused by site conditions and modifications to a project that take place after the actual construction on site has already started leads to discontinuities between the design and execution of the project.
- **Clients and some stakeholders such as local authorities, residents, etc. may have different and wrong perceptions because of the lack of understanding of 2D architectural and engineering drawings or vice versa:** The design team cannot fully understand the client needs

because of lack of communication, a shared platform and a virtual reality (VR) tool understandable to both client and architecture. In the end, these constraints bring about client dissatisfaction.

It is recognised that greater benefits can be obtained and the above constraints can be considerably reduced if a complete integration based on VR tools is achieved. In this respect, the following are considered major benefits of a desired integrated VR environment:

- Improving the coordination and communication among the client, design team members and construction professionals by using standard formats and intuitive VR tools that ease communications and information sharing
- Evaluating the design at very early stages of the project's life cycle, in terms of architectural, technical, financial and environmental aspects since VR tools allow the design team to have a quick and high-quality feedback on the project
- Doing 'what-if' scenarios at the detailed design stage to assess the design solution in lighting, acoustic and thermal aspects
- Closing the gap between the design team and construction team and providing them with an integrated platform in which they can collaborate for the best buildability and applicable construction planning
- Easier use of past project knowledge and information for new developments

Furthermore, visualisation in conjunction with integrated (model-driven) construction systems can be expected to

- enable designers, developers and contractors to use the VR system and virtually test a proposed project before construction actually begins;
- offer 'walk-through' views of the project so that problems can be found and design improvements can be made earlier;
- provide a free flow of information between computer-aided design (CAD) systems and other applications' work packages to minimise misinterpretation between project participants;
- facilitate the selection of alternative designs by allowing different plans to be tested in the same virtual world.

The development of the CIE systems as a building information modelling (BIM) in Case Studies 3 and 4 demonstrates that these systems can meet the above needs of the construction industry. The

CIE systems in those case studies are already explained in Chapter 3. Therefore, the main focus of this chapter will be on formulising the RE process and its implementation in the case study projects.

6.3 The Requirements Engineering Process

It is hard to find well-formulised methods to support the entire design of CIEs. Initial RE activities were not well planned in Case Study 3. Rather, they evolved as the team learned more about the CIEs. Since the group members were from different backgrounds, there was no shared understanding and agreement on the RE strategies.

Initial activities of requirements capture and modelling were not regarded as a process that requires interaction and a shared understanding between the user groups and the developers. At the outset, there were some isolated requirements capture activities amongst the user groups, particularly in Case Study 3. Hence, a shared understanding between the end-users did not emerge at the outset. This lack of shared understanding reduced the effectiveness of the requirements activities. The RE activities undertaken in Case Studies 3 and 4 are explained below.

6.3.1 Use case modelling

Unified modelling language (UML) provides object-oriented development and fits well with some other technical aspects such as Industry Foundation Classes (IFC). The technical team also supported the use of the UML for RE activities. Therefore, use case-driven analysis was selected for requirements elicitation and analysis. Rational Rose Enterprise Edition software tool of UML was selected for the development of high-level use cases and for the decomposition of these use cases into further detailed object diagrams before committing them to code.

Use cases that outlined the CIE systems in Case Studies 3 and 4 were developed before delving into the details. For example, in Case Study 3 in the early iterations, each application of the CIE system was considered as a high-level package use case, which is considered architecturally significant and described in detail. The detailed descriptions of the use cases were done in the flow of events, on which the stakeholders agreed. In this front-end activity, the use cases were used to capture what the system should do from the user's point of view. Thus, use cases act as the common language for communication between the end-users and the developers.

6.3.2 Contextual design technique

The technical team in Case Study 3 thought that the use-case diagrams developed were of high level, not scaled well with clear boundaries, and they were not descriptive and sufficient enough for the developers to implement them. Therefore, the CD technique was used to develop the requirements specifications. Through close collaboration with the industrial partners in the user group of Case Study 3, some useful outcomes were produced, such as the requirements specifications, by means of the CD technique for the technical partners to start the development with.

6.3.3 Storyboarding for acquiring tacit knowledge

Storyboarding was taken into consideration later in Case Study 3. The user group held a workshop, which resulted in the emergence of a shared understanding amongst the end-users. The user group focused on further details, such as how the CIE system should function and how the stakeholders such as the client, architect, engineer and contractor should act to conduct their duties using the relevant functions of the CIE system. Fifteen scenes were developed and in each scene the duties of the construction stakeholders and how they should interact to conduct their duties collaboratively were explained. Storyboarding enabled the technical team to understand the requirements clearly and the shared understanding was extended amongst the technical team and the end-users. Since the initial RE efforts were actually an experimental learning process, these efforts were chaotic. However, subsequent to the workshop, the RE activities became relatively more structured and turned into a shared process, and a shared understanding evolved.

Since the storyboarding enabled a shared understanding amongst the project team, it was further developed to be used in the testing process. The requirements specification was updated through storyboarding during the system tests. Furthermore, storyboarding enabled collaboration between the members of the user group.

Storyboards can be used in the early stage of the system development life cycle for capturing the user requirements, whilst use cases developed during the requirements capture can be of further value during the analysis and design activities, also playing a significant role in the testing process (Leffingwell and Widrig, 2000). Storyboards could be considered as a dynamic format of the use case (Hurlbut, 1997) and are a more structured format for scenarios than text-based use cases. They can also act as a high-level use case and be utilised as test cases, which was done in Case Study 3 where storyboarding evolved throughout the testing phase.

6.3.4 Incremental prototyping with the user tests

The testing process started when the first prototype was released and continued until the completion of the project. Before the release of the initial prototype, some preliminary preparation for the tests had been done. For example, a test plan and a testing strategy were developed. The responsibility and duties of each member in the user group were designated in the test plan, based on an agreed consensus between the members of the user group. Consequently, a process was defined, and a shared understanding between the end-users evolved at this stage.

In the development of CIE systems, the first thing that is very important is to have early prototypes to demonstrate to the end-users. Therefore, the released early prototypes underwent black-box testing conducted by the user group. A well-documented strategy for the testing process was identified. In addition to this, the well-structured testing documentation enabled the user groups to envisage how to contribute to the testing process and interact and collaborate with each other throughout the testing process. From the very beginning of the testing process, both universities in the user group worked very closely, leading the users throughout the process. The testing process played a major role in evolving the requirements and providing a spiral development process.

In Case Study 3, the testing process incorporates three main testing stages: alpha, beta and final testing. The tests were also called *black-box tests*. This is because the tests evaluate the programs against the written requirements specifications and observe the programs as black boxes, and are totally unconcerned with the internal structure of the programs (Lewis, 2000). Furthermore, there was a symbiotic relationship between the end-users and the developers in the project. The end-users conducted all tests in the life cycle. The developers released the prototypes and the end-users, who were distributed across Europe, tested the prototypes continuously in a collaborative manner. They continuously provided feedback to the developers throughout the test phases.

The most effective way to reduce risk is to start testing early in the development cycle and to test iteratively with every build. With this approach, defects are removed as the features are implemented. Each testing phase – alpha, beta and final – is conducted in an interrelated, continuous and iterative manner. Contemporarily, it is recognised that for almost all systems, it is right and necessary to have some kind of iterative process. Modern development processes consider iteration as fundamental and try to provide ways of managing, rather than ignoring, the risks (Stevens and Pooley, 2000).

Regarding the spiral process, it provides as many iterations as required. However, it is noticeable that the spiral process does not prescribe how each phase is done. The ambiguity about what is meant by iteration is one of the main differences between modern technologies, which have their own versions of the basic spiral (Stevens and Pooley, 2000). Therefore, the user group employed the UML test workflow to conduct the tests in a stage-wise manner. A typical UML workflow encompasses the stages: plan test, design test, implement test and execute test in the integration tests stage and execute tests in the system test stage.

Plan test

The purpose was to identify and describe the testing that would be implemented and executed. Initially, a single test plan was developed, describing all the types of tests to be executed for successive early releases of the CIE system. The test plan was modified for each testing stage. On the basis of the test plan, the test efforts were measured and managed.

Design test

The purpose was to identify, describe and generate the test model. In the design activity, the test designer analysed the target of the test and developed the test model. The test design transformed the requirements specifications into alpha, beta and final test cases. In Case Study 3, the outputs of use case development, CD and storyboarding were utilised for the test design. For example, the use cases and storyboard drove the design of the software elements that implemented the tests.

Implement test (alpha phase)

At this stage alpha was defined as 'a software package capable of demonstrating the concept'. Testing of the alpha version primarily included the building of a full-scale demonstration based on the chalet case study and the feedback obtained from the demonstration. The main elements of the alpha phase testing were the functionality and usability testing. Each product's conformity was observed and examined against the user requirements through functionality testing.

At the alpha testing phase, six applications (Client Briefing, Lighting Simulation, Acoustic Simulation, Thermal Dimulation, 4D Simulation and Site Planning and Analysis), were tested as stand-alone applications in the United Kingdom, France, Denmark and Finland at the same period of time. At this stage, predetermined user requirements captured for each application were used to test the products. User requirements were described as the wish list of the end-users, which denoted the functionalities of the applications. The advantage of alpha testing was that it enabled user groups to test the individual components and explore the possible defects belonging to different applications and log them. This provides a better way to manage integration of all products at the integration phase.

Execute test in the integration test stage (beta phase)

At this stage, beta is defined as 'a software package that can be run by the end-users in a session guided by the developers'. A design of an office building provided by the Finnish partner of the project is the case study for the beta testing. The main focus of beta testing is the integration for exchange of information through the communication layer between the stakeholders situated in discrete places, which corresponds to integration and distribution testing.

The initial storyboard was further developed and transformed into a test case tool to drive the integration testing for the CIE systems in both Case Studies 3 and 4. The storyboard, which was related to the construction supply chain, enabled complete retest of all the functionalities of each product in order to observe the expected improvements between the alpha and beta tests in functionality. Lastly, the defects encountered at the alpha-testing phase underwent regression testing at the beta phase in which the defects have been supposedly removed. Furthermore, new defects discovered at the beta phase were also recorded.

Execute test in system test stage (final phase)

Final phase refers to the system testing that is a multifaceted test to assess the operation of the system in functionality, reliability, performance and usability. System testing tests the collection of the applications that constitute a deliverable system, and the entire domain is considered to satisfy the criteria for a system test (Whittaker, 2000). At this stage final testing is defined as 'a software package that can be run independently by the end-user'.

At the final phase, the storyboard was modified from the beta version and it required interaction of the applications of the CIE system. That is to say, the storyboard at this stage corresponds to a higher level of the construction supply chain. Therefore, it entailed the interactions of the applications and the full usage of the CIE system. The final case study was again the office building used in the beta phase. Besides, the issues tested at the beta phase were retested under regression testing.

Acceptance testing was the end-user reaction to the system and their evaluations on how much the system impacts their business life. To ensure the acceptance from the end-users, industrialists were involved in the integration and system testing stages. Case Studies 3 and 4 had what the industry wants in real construction life via these RE activities.

6.4 The Requirements Deliverables from Use Case Modelling

Case Study 3 initially used use cases and UML as the means of requirements capture. However, as discussed in Chapter 2, use case approach was by itself proved inadequate to capture the diverse user requirements in software development. At the earlier stages of the projects, RE activities were conducted separately amongst the user groups of the project team. For example, some used use case driven requirements capture; the others exploited the CD technique for requirements capture. There was no link established between the two methodologies for a better requirements capturing.

The requirements workflow proposed by Rational Unified Process (RUP) was used to communicate the complex top-level industry requirements and progressively break into more detail. Furthermore, to manage the large-scale and evolving industry requirements, Boehm *et al.*'s (1998) recommendations for a 'win–win spiral model' were also taken on board, which indicated creating three critical milestones, i.e. (i) life cycle objectives, (ii) life cycle architectures and (iii) operational capability. A number of techniques were used to communicate the complex top-level industry requirements and progressively break into more detail. In particular, the following approaches from RUP were adopted:

- Vision statement
- Stakeholder perspectives
- Use cases
- Systems requirements

6.4.1 The vision statement

A major review was conducted by Sarshar *et al.* (2000) to summarise the research and industrial best practices efforts in the form of a vision for the next decades of construction projects and the way their processes will be supported by IT. The purpose of this vision was to (i) paint a picture for the industry, (ii) provide a platform for discussion and debate, (iii) serve as a platform to identify gaps between research and industrial uptake, (iv) assist in identifying barriers to vision realisation and (v) illustrate how IT could benefit the industry. This vision was not perceived as complete and final. Rather, it was a live and dynamic picture that sets the scene for future developments. The vision identified seven key themes in research and development in CIEs as follows:

1. Model-driven, as opposed to document-driven, information management on projects
2. Life cycle thinking and seamless transition of information and process between life cycle phases
3. Use of past project knowledge/information in new developments
4. Dramatic changes in procurement philosophies, as a result of the Internet
5. Improved communications at all life cycle phases, through visualisation
6. Increased opportunities for simulation and what-if analysis
7. Increased capabilities for change management and process improvement

In Case Study 4, on the other hand, vision development was carried out through subsequent national and international workshops. Many people from industry and academia attended these vision development workshops, which identified a clear and strong vision for the future and also requirements and their prioritisation in the CIE development in Case Study 4. The themes in these workshops were similar to those mentioned in the earlier list, and also include identifying and prioritising key requirements for the CIE development.

6.4.2 Stakeholders' perspective

The construction industry has a number of major stakeholders, each of whom have a different and sometimes conflicting perspective. The stakeholders include (i) clients, (ii) architects, (iii) contractors, (iv) subcontractors and (v) engineers. It was important to have the perspective of each stakeholder in view. Case Studies 3 and 4 captured the requirements of each group as a starting point to understand the industry requirements. The industrialist members of the user group put forth their own perspective. These perspectives provided high-level requirements and a baseline for future developments. The policy in Case Studies 3 and 4 was to be client led. Where there is a conflict of interest between the different perspectives, client requirements will take precedence.

6.4.3 Use case modelling

The vision and the stakeholder perspectives provided the life cycle objectives for the industry and for Case Studies 3 and 4 projects, respectively. Top-level architectural diagrams and some functional decomposition diagrams were developed to define the scope, for example, in Case Study 3 and demonstrate how it fits into the larger industry requirements. Case studies 3 and 4 did not have sufficient resources to develop systems for all aspects of the construction industry and construction life cycle. For example, in Case Study 3, the areas within the project scope included client briefing, conceptual design, detailed design and building construction. At this point, hierarchical use cases are developed for each of these four areas, as illustrated in Figure 6.1.

The project strategy in Case Studies 3 and 4 was to select one use case for initial development and implementation. On the basis of learning from this initial use case, the rest were rolled out. The initial use case was detailed design. The rationale for this choice is that (i) the implementation requires input from all the partners and (ii) it includes many of the CIE systems' core functionalities in both Case Studies 3 and 4. The use cases were checked with the industrial partners. Generally at this high level, the requirements were similar among all the countries. It is evident that at a more detailed process and procedure level the working practices vary between different countries, and this led to the decision that at a technical level, appropriate plug-ins could be developed for each country.

The use cases have, in general, proved successful in capturing the life cycle objective. They have provided a platform for discussing the project scope, and identifying missing elements in the requirements.

6.4.4 Systems requirements: high-level technical requirements

Since Case Studies 3 and 4 are an evolving environment, the projects decided to define top-level technical requirements from the users'

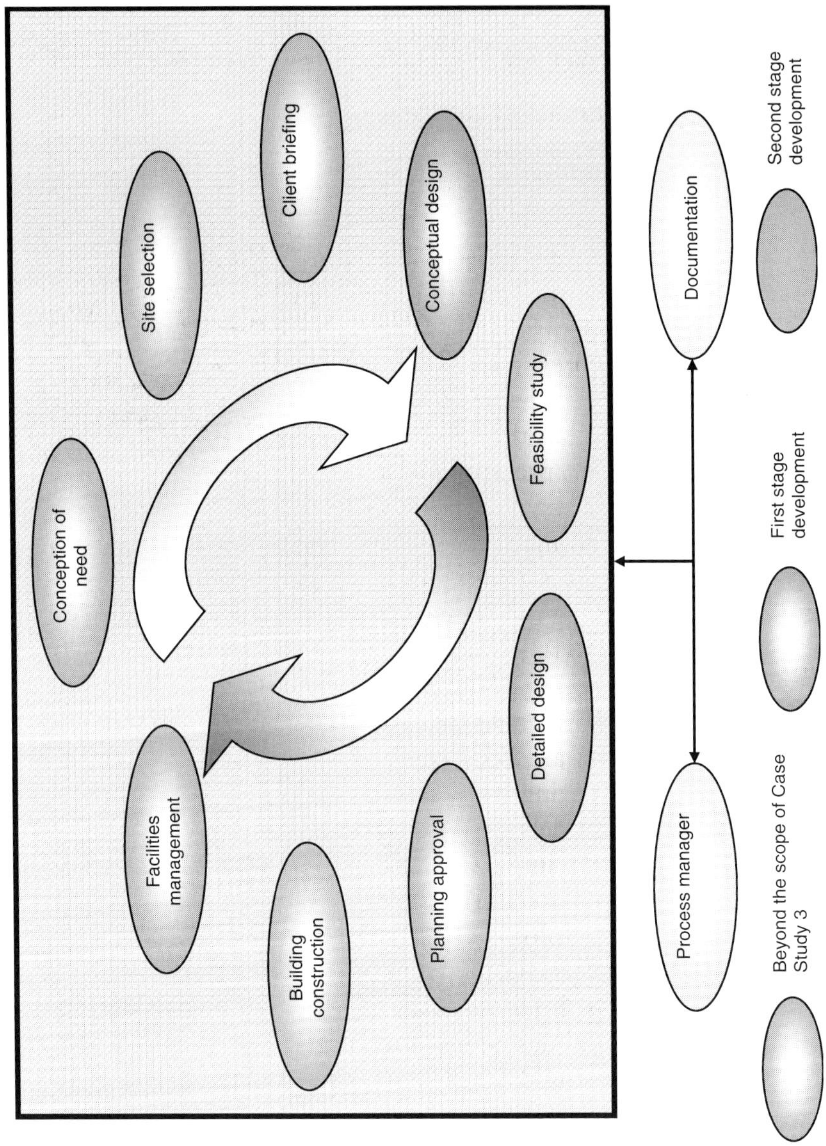

Figure 6.1 Use case diagram for construction life cycle.

perspective. It is a short description of (i) the hardware platforms the industrial users prefer, (ii) the need for the CIE systems to interact with other construction applications and (iii) the users' perception of how the system should work.

6.5 The Requirements Deliverables from Contextual Design Technique

Often, requirements modelling starts with UML based on object modelling of the system with efficient tools such as Rational Rose (http://www.rational.com). This results in giving too little consideration to understand user needs and user interface design as well as quality requirements. The risk was obvious that important properties of the final product with regard to end use are overlooked (Christiansson *et al.*, 2001). There were not so many well-formulised methods to support the entire design of a CIE system such as the one in Case Study 3. Owing to its well worked out user-centred approach, in Case Study 3, the CD method was chosen to take into account the end-user work practice and interface requirements.

The lack of synthesis between use cases to capture system usage aspects is the main drawback of use case modelling design. Hence, employing the CD approach before use case modelling could bring about a strong synthesis between use cases to eliminate the possible risks mentioned above. However, the CD method was partially implemented because there was no agreement on the use of this technique between the user teams of the project. For example, a complete contextual inquiry process and the incremental progress refining and consolidating the system form and functions (structure and time-dependent models) could not be conducted. However, it was important to design and implement efficient collaboration and documentation tools.

CD methodology was used by the user group of Case Study 3 in collaboration with the industrialists. As explained in Chapter 3, CD has five different types of work models at the vision phase of the contextual design methodology. The following work models were developed using the CD approach.

- Artefact models for lighting, acoustic and thermal design
- Top-level sequence models for lighting, acoustic and thermal design
- Top-level workflow models for lighting, acoustic and thermal design

The above models for lighting design are shown in Table 6.1 and in Figures 6.2–6.5. Sequence models show design intent and the

Table 6.1 Artefact modelling for lighting design.

Action	Artefacts
Access permission/demand	Contract/agreement
Get geometry details	Drawings (electronic, hard copy)
■ Room type/class	Text documents (electronic, hard copy)
■ Surfaces	Telephone, email program
Idea generation	Best practice, previous experience (intranet DB, hard copy), supplier information, new developments
Define light principles	Best practice, previous experience (intranet DB, hard copy), supplier information (catalogues, telephone, email, Internet)
Set up requirements from external constraints	Health regulative/standards
Light geometry	Standards/codes
■ Maintenance factor	Built-in third party tools
■ Working level	Drawings/documents from architect
■ Mounting height	Telephone, email program

Adapted from Christiansson *et al.*, 2001.

workflow models show how these intents are achieved. The sequence models are complemented by the artefact models, which show how the design artefacts are manipulated and with what tools. They also help to reveal the design intent and what the team, groups and individuals think about their work. Real data and cases were used during the construction of the artefact model. The artefacts support communication as in the following:

- Artefact group in relation to intended and/or real exploitation (supplier information, requirements from authorities, other public information, communication tools, firewall analysis and simulation programs, product and process models and process manager)
- Artefact properties and characteristics (personal/shared, Case-Study 3-specific/general, synchronous usage, access rights, access levels, artefact memory, alternative artefacts for the same activity, alternative virtual workspace activities with use of the same artefact, identification icon and name)

The flow and sequence models were combined with artefact models and synthesised to storyboards, as illustrated in Figures 6.5 and 6.6. Using storyboards, the team developed the vision into definition of

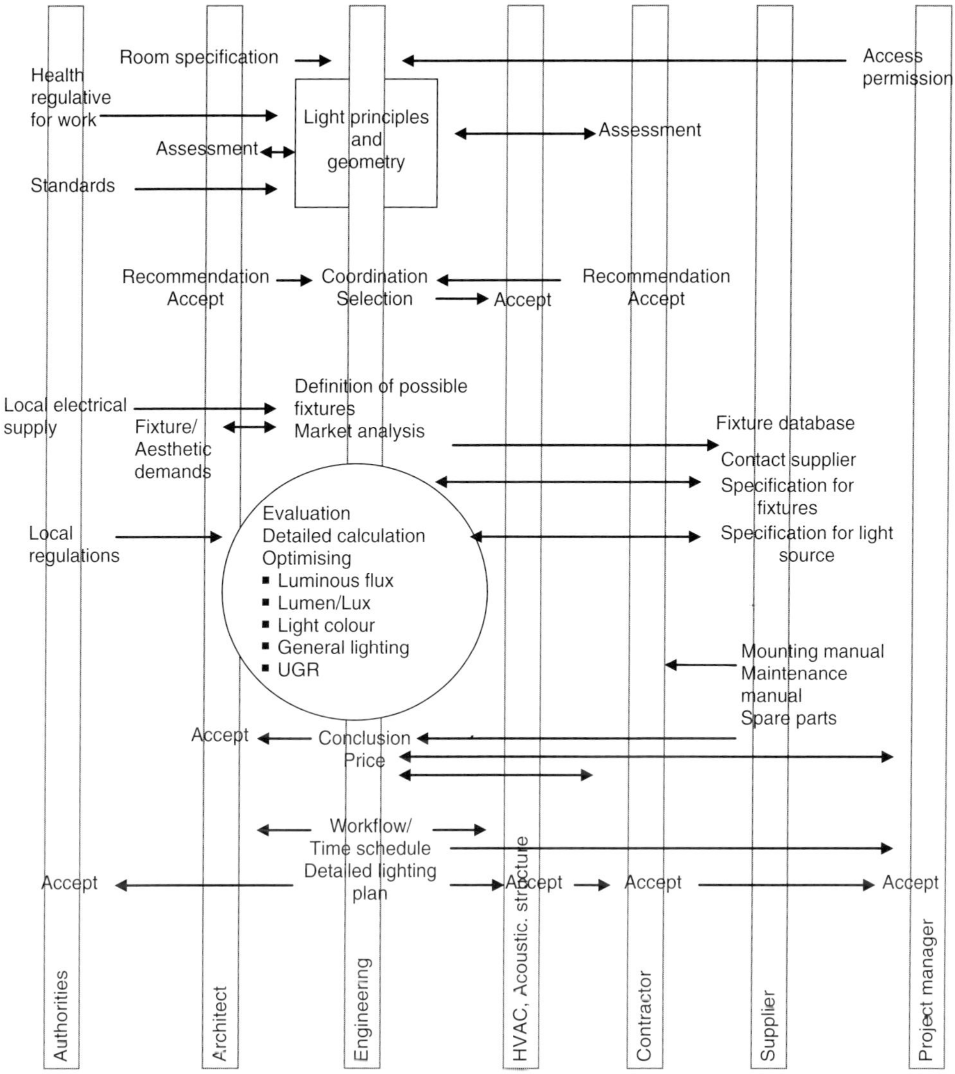

Figure 6.2 Consolidated sequence model for detailed lighting design. (Adapted from Christiansson *et al.*, 2001.)

how people would work in the new system and ensured that all aspects of work captured in the work models were accounted for. Afterwards, it was time to implement the user environment design (UED) with detailed user-interface proposal. It does not prescribe the order of work, which is valid for storyboarding. Objects and other knowledge representations were further specified to meet user-induced requirements. Figure 2.3 in Chapter 2 shows the progression from design to development.

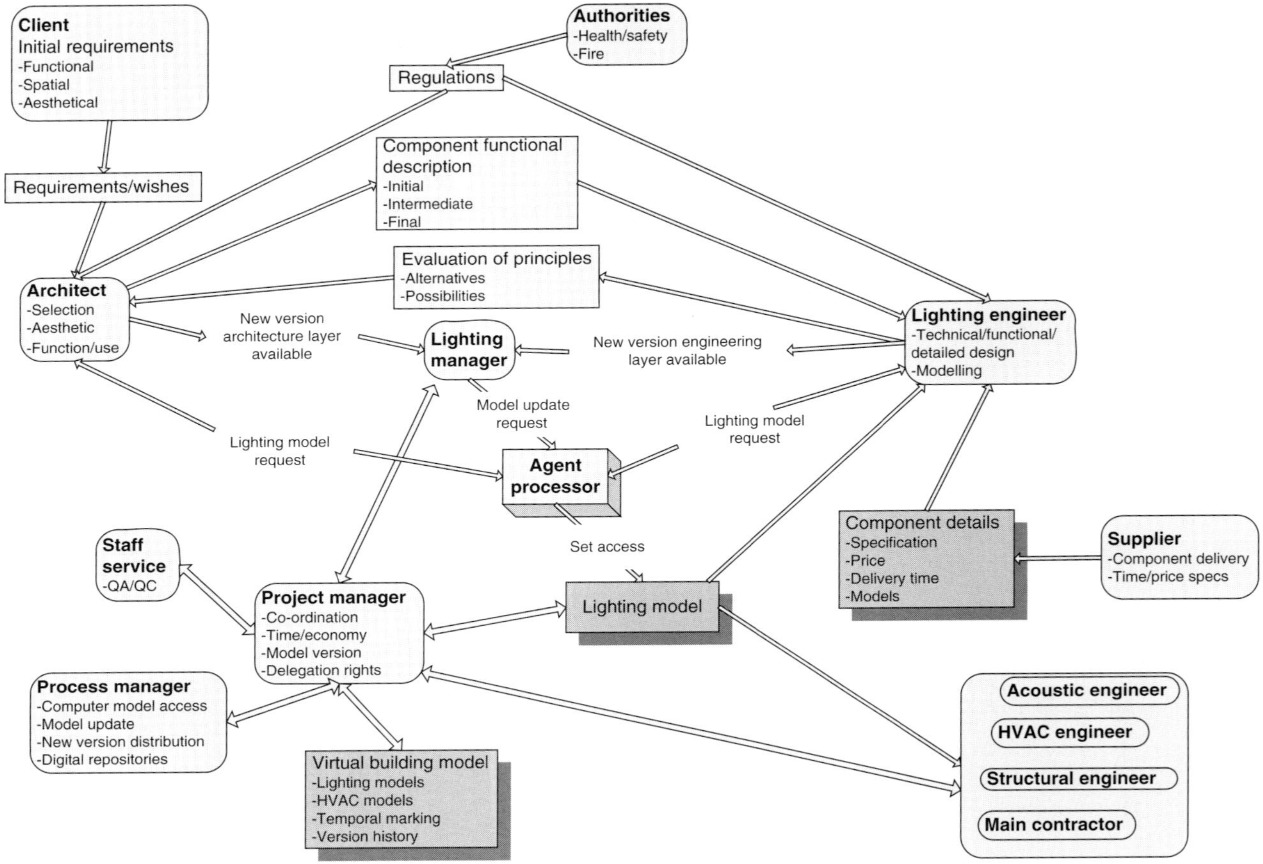

Figure 6.3 Part of top-level workflow lighting model from many simultaneous individual perspectives. (Adapted from Christiansson *et al.*, 2001.)

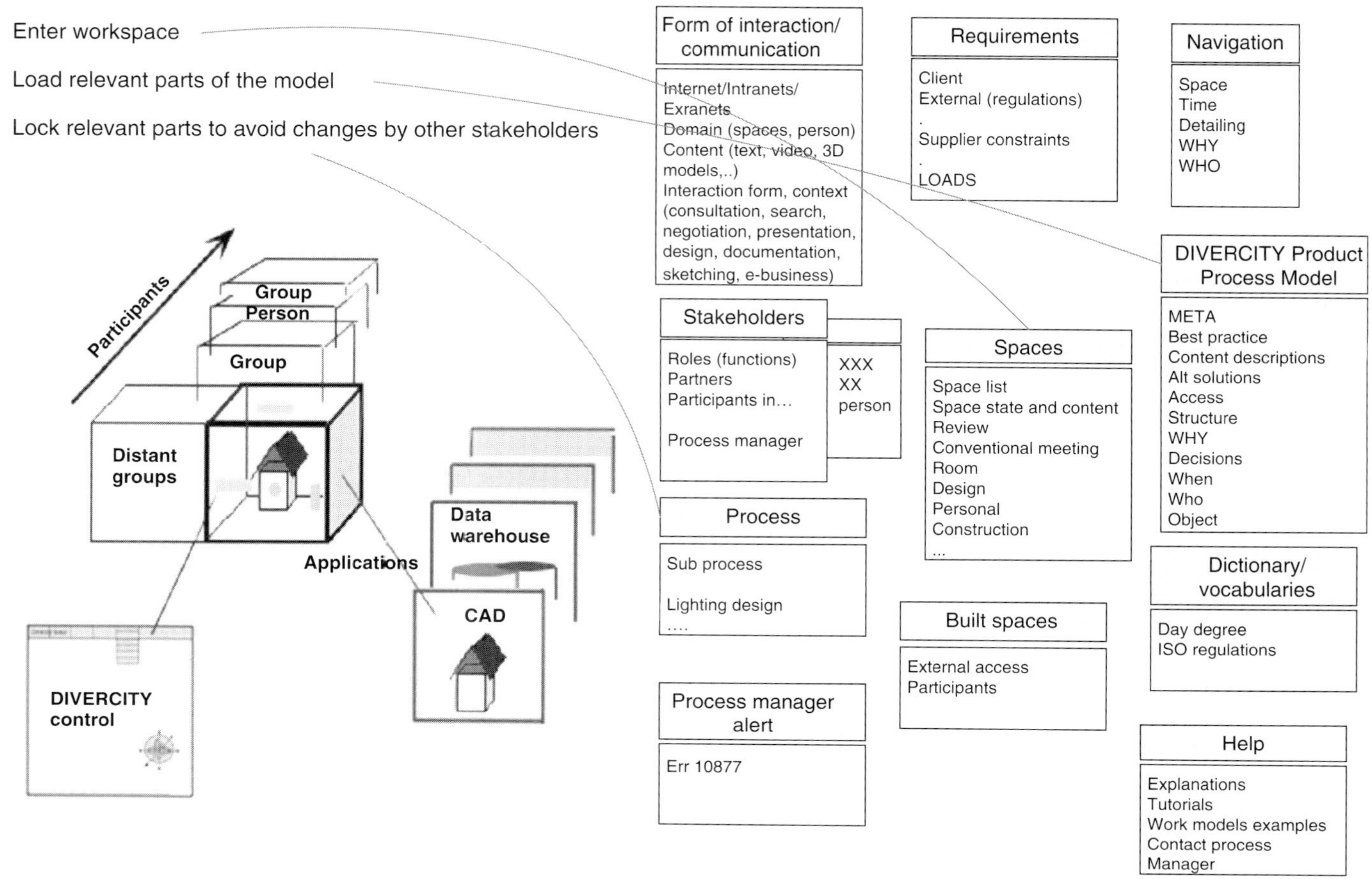

Figure 6.4 High-level storyboard for lighting simulation during design. (Adapted from Christiansson *et al.*, 2001.)

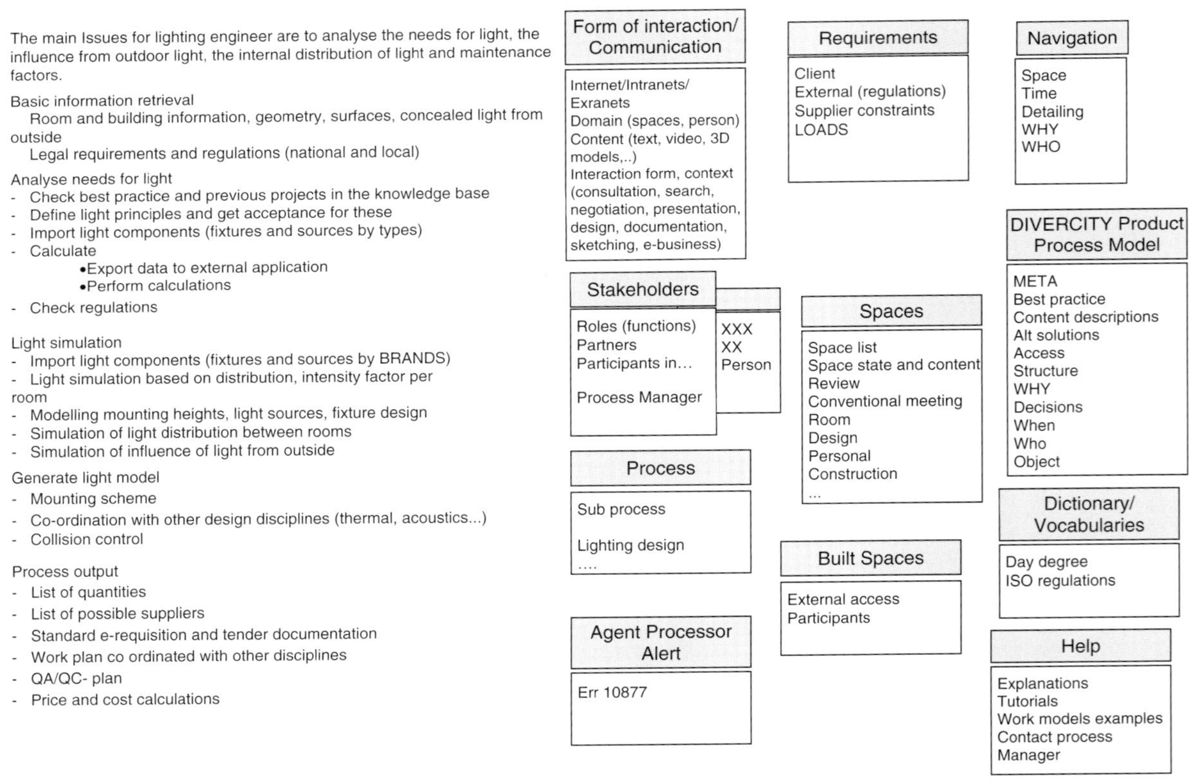

Figure 6.5 Storyboard in Figure 6.4 is further detailed before transition to user environment design. Artifacts on the right are used to fulfil the functions on the left in the sequence shown. (Adapted from Christiansson *et al.*, 2001.)

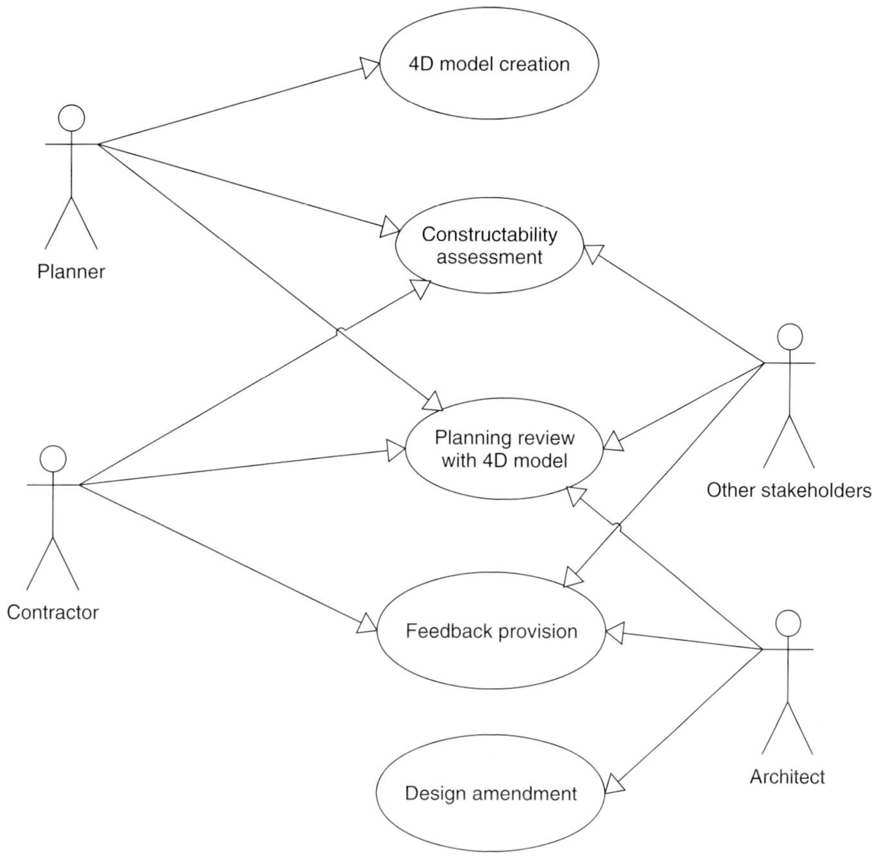

Figure 6.6 A simple use case diagram for construction planning and constructability.

6.6 The Requirements Deliverables from the Incremental Prototyping with the User Tests

In Case Studies 3 and 4, continual validation testing was used as part of the incremental approach to software development. That is to say, testing was not comprehended as a single activity nor was it at a phase in the project during which the quality was assessed. The technical partners obtained continuous feedback on the evolving system functionality and quality.

The user group tested the broad functionality of early CIE systems and the stability, coverage and performance of the architecture while there was still opportunity and sufficient time to fix it. Furthermore, the user team tested and validated the final product to assess its readiness for delivery to end-users, which is called *acceptance testing* in the software testing literature.

6.6.1 The testing methodology in Case Studies 3 and 4

Conformance and interoperability testing are two different approaches to validating a software's usefulness with respect to a specification. The main difference between the two is that the traditional conformance testing approach compares an implementation against written requirements specifications, whereas the interoperability approach compares one implementation with others.

The testing approach was conformance testing, by which the CIE systems in both Case Studies 3 and 4 and their components were compared with the user requirements, which had been already determined by the user groups of the project team through interaction with industrialists, academicians and IT specialists. The conformance approach was suitable and provided some advantages. For example, an advantage of conformance was that each application was compared to the same requirements specification. The breadth of coverage of the user requirements could be measured and redundant tests could be minimised.

On the other hand, the conformance testing approach has been criticised, as it may overlook tests that are important in real world implementation and it can catch things that are irrelevant in the real world. However, in the development of innovative systems such as the CIE systems, user requirements are captured by envisioning the use of the system. This envisioning can cause catching the irrelevant things in the real world. In Case Studies 3 and 4, this problem was overcome by holding user-testing workshops with industrialists, who would provide their views on the applicability of the systems to the real world.

6.6.2 Use cases and the storyboard in the user tests

A use case can be viewed as an object and likewise the test can be viewed as an object. A test model incorporates the test specifications, which are tightly related to use cases and the results of the testing. Test specifications can be viewed as a test's class and thus common parts of test specifications (use cases) can be decomposed into several test specifications. A test execution is an instance of a test specification. The instance quite clearly has behaviour and also attributes. The outcome of a test execution is a test result.

Each use case describes a requirement on the system so that a correct system design allows each use case to be carried out. An obvious and very useful technique for validating a system design is to take each use case in turn, and where a use case includes significantly different families of scenarios, an example of each use family should be included.

In the storyboard developed in Case Study 3, a number of use cases guided the functionality tests. The storyboard was modified and re-employed in each test phase. This resulted in effective and continuous re-tests of the applications in many respects, such as functionality, usability, performance and reliability. A part of the storyboard from Case Study 3 is given below as a use case diagram (Figure 6.6). The flow of events of the use case for construction planning and constructability in the relevant scene of the storyboard, which was demonstrated by the CIE system in Case Study 3, is as follows:

The planner will

- obtain 3D building model over the distribution module;
- schedule construction generic tasks in MS project using the documentation provided by design team members;
- invoke 3D IFC model and the building schedule into 4D Construction Planning Simulation;
- link the IFC model with the scheduled tasks;
- create/save 4D model in 4D Construction Planning Simulation;
- convert 4D IFC model into VRML format;
- experience interactive visualisation and walk-through in 4D simulation to observe constructability; in terms of construction activities and methods, required resources, time and cost;
- revise construction plan;
- upload the 4D model onto distribution module.

The designer, contractor, consultant, subcontractor and supplier will

- import the 4D model from the distribution module;
- experience 4D simulation via interactive visualisation and walk-through;
- review construction plan and assess constructability from their perspective.

If any of the above stakeholders notice a constructability problem in design, which that affects constructability badly, the following will be done:

The designer will

- review the design with the stakeholders using 4D Construction Planning Simulation in order to amend the design for a good constructability;
- review the design and show the amendments in design by 3D visualisation to the client;
- provide the amended design to the construction management team.

The contractor, consultant, subcontractor and supplier will

- provide the planner with feedback according to the latest changes in order to update the construction plan for a better constructability over the distribution module.

The construction planner will

- finalise the construction plan based on the feedback obtained from the stakeholders;
- upload the 4D model onto distribution module for the site planner.

The site planner will start to organise the site layout for the actual construction life cycle defined in the construction plan.

6.6.3 Testing results

The testing results were useful to both the end-users and the technical team. To be precise, they provided feedback to the technical teams to improve the system and provided a better vision and understanding to the end-users to improve the user requirements specifications. As a result, the testing results provided what they were supposed to.

Rational Rose software as a testing tool could have been used for more effective testing. However, this would involve a learning curve for testing activities to be efficient. Furthermore, another ambiguity was whether Rational Rose tool enabled the end-users to undertake the testing in collaboration with a shared understanding.

6.7 Critical Analysis and Reflections of the Requirements Engineering in DIVERCITY

This section summarises the critical aspects, reflections, benefits and lessons, learned from requirements engineering experience in Case Studies 3 and 4. These are listed below:

- Initially, owing to the lack of a strategy for requirements elicitation and analysis, the user group could not work as a team and worked separately by using different requirements approaches.
- Owing to a lack of strategy determined for the RE, the user group did not make any evaluation and assessment to select the right methodologies that best suit the nature of the CIE systems development, such as Case Study 3.
- The requirements capture process was not understood and shared by the whole user team.

- None of the end-users in the user team had a background in RE, which was desperately needed.
- Although there was a consensus to purchase the Rational Rose software for RE, the learning curve of the Rational Rose software took a vast amount of time, within which a significant use case development for the technical team could not be done.
- The lack of a shared understanding between the end-users in the user team reduced the quality of the RE effort.
- The requirements activities turned into a process during the storyboarding and the prototyping, which resulted in the emergence and evolution of a shared understanding amongst the end-users.

The benefits of the shared understanding were as follows:

- Testing process emphasised the continuous improvement in both requirements and software development.
- Individual requirements efforts were gathered and became teamwork based, which helped the user team to improve the requirements specification and furthermore, the user team validated the requirements through testing.
- In addition, the continuous testing process permitted the system to be user driven.
- The user tests were not only done for verification and validation of the system, but they were also done to introduce it to the end-users who are not very familiar with innovative computer integrated construction.
- A process of workflow called *storyboard* was derived from the user requirements, which implies a construction supply chain for a lean construction work environment. All tests were conducted according to the storyboard by the end-users and not the developers. On the basis of the consecutive test results, the developers released subsequent prototypes until the end-users were satisfied in the defined test criteria in every phase of the testing process.
- The use of storyboard was of significant importance in a problem domain where the scope is very broad and ill-understood. In Case Studies 3 and 4, the applications are pushing the boundaries of current construction industry practices. The industry was not clear on the requirements and applicability of the novel products during the early stages of the research. The storyboard and the subsequent more detailed scenarios played a major part in requirements capture as well as the testing in both Case Studies 3 and 4.
- The tests were undertaken in different countries at the same period in time in a distributed manner, which brought an extra novelty

to the testing approach but also brought about extra difficulty in managing the test activities. Because the test teams were geographically distributed, the test plans were communicated and a consensus reached prior to and during every testing phase. Under this circumstance, communication and interaction between the testers distributed across Europe were extremely important for effective tests. Another dimension of the difficulty of distributed testing was that test teams in different countries conducted the tests independently, which resulted in different interpretations in discrete countries. These various interpretations were required to be harmonised and a shared interpretation of test results needed to be provided to the developers.
- The business strategy for marketing and exploitation were executed more effectively because of the user tests. To be precise, business strategy and exploitation were not clear enough before the tests started. The user tests in Case Studies 3 and 4 enabled the project team to clearly define the business strategy for marketing and exploitation of the system.
- Shared understanding evolved over time; therefore, the test results became increasingly more coherent.

Some reflections that imply further improvement in the RE approaches of Case Studies 3 and 4 are addressed below:

- Initially, the research team should have looked at evaluation of the RE methodologies and the right methods should have been identified for the CIE system. There was no consensus established in the user group on the correct RE techniques in Case Study 3. The most suitable techniques to enable the establishment of a process framework for RE process could have been selected by comparing several state-of-the-art requirements methodologies to find out the strengths and weaknesses of each methodology. Accordingly, the project team would have selected the right RE techniques by establishing a consensus and subsequently a shared understanding amongst team members would have emerged.
- It would have been more useful if the use cases were developed after the requirements elicitation according to the CD technique, which normally provides a coherent synthesis with use cases and enables the requirements analysts to establish a solid basis for the use case development.
- The testing process should have been started 6 or 9 months earlier.
- The testing process did not include piloting, which could have been a considerable progress in the implementation of CIE systems

from Case Study 3, and Case Study 4 could have contributed to overcoming the barriers to CIEs' implementation in construction.

- The user requirements capture in Case Studies 3 and 4 had insufficient links with marketing requirements. That is to say, these visual workspaces can be developed but they will never be taken on board by the construction industry because the marketing aspect has not been looked at. Strategically, first of all, requirements specification should be market driven and secondly, the industry barriers against the implementation should be taken into account in the requirements capture process continuously. The two critical aspects above were not taken into account in the whole RE activities. This is because in Case Studies 3 and 4, piloting the system on real construction projects could not occur because of resource constraints although there were wishes to exploit the system on real commercial construction projects together with industrialists. The piloting, which should have been incorporated into the testing process, could have been a very good solution to the two critical drawbacks of implementation above.

References

Boehm, B., Egyed, A., Kwan, J., Port, D., Shah, A., Madachy, R., (1998), "Using the Win Win Spiral Model: A Case Study", IEEE Computer, 7:33–44.

Christiansson, P., Svidt, K., Skjærbæk, J.O., Aaholm, R., (2001), "User Requirements Modelling in Design of Collaborative Virtual Reality Design Systems", Mpumalanga, South Africa, International Conference on Construction Information Technology.

Hurlbut, R., (1997), "A Survey Approach for Describing and Formalising Use Cases", XPT-TR-97-03, http://www.iit.edu/~rhurlbut/xpt-tr-97-03.html.

Leffingwell, D., Widrig, D., (2000), "Managing Software Requirements", Essex, England, Addison Wesley Longman, Inc.

Lewis, W.E., (2000), "Software Testing and Continuous Quality Improvement", Plano, Texas, Software Testing Associates Inc.

Sarshar, M., Betts, M., Abbott, C., Aouad, G., (2000), "A Vision for Construction IT 2005–2010", London, United Kingdom, RICS (Royal Institute of Chartered Surveyors) Research Series, pp 1–42.

Stevens, P., Pooley, R., (2000), "Using UML Software Engineering with Objects and Components", Essex, England, Updated Edition Pearson Education Limited.

Whittaker, J.A., (2000), "What is Software Testing? And Why is it so Hard?", IEEE Software, 17(1):70–79.

7

Evaluation of the Requirements Engineering Practices

7.1 Introduction

The software development world is experiencing an irreversible trend towards the globalisation of business, a phenomenon that is rapidly gaining the attention of the research community (Damain, 2003). Seeking lower costs and access to a global resource pool are the main factors that accelerated this trend. As a consequence, requirements engineering (RE) processes in which the communication between stakeholders, in particular decision-making, is taking place across geographical boundaries, and is becoming crucially important in the development of the integrated, distributed and collaborative software systems such as the computer integrated environment (CIE) systems for the construction industry.

It is asserted that the CIE systems can provide many benefits to multidisciplinary sectors such as the construction industry. These benefits are already elaborated on in Chapters 3 and 4. Since it is also asserted that requirements capture is an issue that can contribute to the successful implementation of the CIE systems, it is necessary to analyse and evaluate RE implementation in the case study projects, in particular, the CIE developments in Case Studies 3 and 4. This would help in curing the weaknesses and problems and then enhancing the RE process before proposing a constructed and ready-to-use RE method for both commercial and research-based future developments of the CIE systems for the multidisciplinary sectors, such as construction.

In order to point out the strengths and weaknesses in the RE experiences, this chapter will evaluate and assess from multiple viewpoints, the RE experience in the case study developments. This evaluation is mainly based on the evaluation criteria identified in Chapter 5, which enable a comprehensive evaluation in many perspectives at organisational, project, business and professional levels. In Section 7.2, the scope of the evaluation and assessment model is described.

In Section 7.3, the methodology of the quantitative analysis and the survey structure are described and lastly, in Section 7.4 a detailed analysis and interpretation of the survey results are represented.

7.2 Scope of the Evaluation and Assessment Model

To make recommendations for improving the RE processes from the case study projects, it is critical to understand the problems experienced and to find out the strengths and weaknesses of the RE process in the project. Whilst in Chapter 6 a critical elaboration of the RE approaches employed in Case Studies 3 and 4 has been carried out including practical difficulties experienced and also the success achieved, in this chapter, the adopted RE process is evaluated more systematically from different perspectives based on a number of criteria defined previously in Chapter 5. These criteria are clustered into five groups, which also reflect the evaluation viewpoints and scope considered. These groups are (i) a match of the CIE system with the construction companies, (ii) user satisfaction and commitment, (iii) the cost–benefit analysis, (iv) the quality of architecture and (v) cost effectiveness of the RE process.

7.2.1 Match of the CIE systems with the construction companies

It is acknowledged that the construction industry lags behind manufacturing in the use of CIE, which is conceptualised as building information modelling (BIM) in the construction industry as a means to increase productivity and efficiency. However, the current work structure is inappropriate for the implementation of the CIE systems as there is no agreed procedural mechanism. This also makes it difficult to compromise a saturated level of fit between the CIE systems and construction organisations.

In developing and implementing new computer systems, some conflicts arise inadvertently because of misunderstandings in communication. Many construction professionals have difficulty in understanding the concepts and terminology used by IT professionals and vice versa. Other conflicts arise from differences in priorities and objectives. Furthermore, in system development, cost and schedule overruns have often been a source of conflict, as well as the conflicts that may be incurred in system development for the construction industry; some forms of resistance may show up in many ways (Paulson, 1995). Most of the reasons for resistance are found in a basic understanding of human nature and social behaviour.

Under these circumstances, RE enables the fit to be established between the CIE systems and the construction organisations. However, there are some drawbacks that reduce the possible high level of fit, such as a lack of process awareness, fragmented nature, cultural attitudes and a lack of familiarity of construction practitioners with the CIE systems. Hence, the end users require training to increase their familiarity with the CIE systems and to address the correct requirements and key issues, etc.

On the other hand, according to the results of the survey conducted by Aouad *et al.* (1998) during the research on the development of construction process protocol (http://www.processprotocol.com/), communications and networking will be a major topic in the future. This will ultimately give the process push development, resulting in some maturity in reaching a consistent, predictable and improved process. As a result, this strategic orientation will increase the fit between the CIE systems and the construction organisations in the near future.

The issues that should be considered are listed in Table 7.1.

Although all the above criteria are required for a successful fit, external problems such as legal issues, fragmented nature, people reluctance, cultural attitudes and so on also result in a poor fit between the CIE systems and the construction organisations. Despite these drawbacks, high-quality RE services in the CIE developments can help to reduce the influence of these drawbacks and compromise the fit between the CIE systems and the organisations.

Table 7.1 Issues for analysis in fitting the CIE system with the construction organisations.

1	The ability of the companies to make the necessary changes to implement the system
2	The match between the CIE system and the strategic orientation of the companies
3	The support from senior management for changes required to implement the CIE system
4	The willingness of the companies to do the required changes to adopt the CIE system
5	The fit between the proposed CIE system architecture and the corporate structure of the companies in construction projects
6	The fit between the CIE system and the technical orientation of the companies

7.2.2 User satisfaction and commitment

User participation and commitment play an important role in dealing with the development of complex systems. Users and developers would benefit from close interaction by exchanging views, identifying and resolving conflicts, as well as sharing information necessary to effectively accomplish the task. The study of the 'participation–satisfaction relationship' hypothesis led to the conclusion that system development 'requires that the appropriate users participate in the project at the appropriate stage and in a manner that enables a meaningful contribution' (Herlea, 1996).

User participation is one of the determinants of system success. The term *user participation* is used when 'referring to the various design-related behaviours and activities that the target users or their representatives perform during the system development process'. The correlation between user participation and user satisfaction reveals the importance of improving user participation within the system development to improve user satisfaction. The criteria to be considered for user satisfaction and commitment are included in Table 7.2.

When these criteria are considered for the CIE systems, it will be realised how important these criteria are for successful development and implementation. For example, awareness of the end users will rise by experiencing the CIE systems, and their prejudice against the CIE systems will partially be weakened. Furthermore, whilst the resistance against the implementation is lessened, the enthusiasm to implement the systems and to defend the CIE concept in different platforms is expected to increase. However, all these expectations can be achieved through the correct RE techniques.

Table 7.2 Issues considered for analysis in user satisfaction and commitment.

1	The level of understanding by the users about what the CIE systems do and do not do
2	The willingness of the users to defend the CIE system against executive management
3	The extent of user consensus on the CIE system implementation
4	Users' awareness about the business changes that are required for the CIE adoption
5	The relationship between the users and the RE staff
6	The user's reaction to the cost estimate
7	The fit between the CIE system architecture and the way the users work
8	Approval of all the documentation by the users
9	The degree of match between the functionality of first release of the system and user expectations

In addition, the approach of IT development in the construction industry has traditionally been technological driven. 'Push strategy' is the most dominant pattern of IT procurement in the construction industry. It has led to a large reported number of failures in utilising IT, which is due to the insensitivity of these systems to organisational design, external environment, and culture and management system (Aouad and Wafai, 2002). Therefore, there has always been a gap between the underlined philosophy of CIE and the environment in which it is aimed to be implemented.

Involvement of the end users in the RE process will lead to a user-driven development, as opposed to one that is technology driven. Subsequently, it results in user satisfaction and commitment to implementation, and furthermore will contribute to bridging the gap between the CIE (or BIM) concept and the work environment in which CIE (or BIM) is to be implemented.

In order to facilitate the accommodation between different stakeholders' perceptions, the RE process should support the involvement of the stakeholders in the formulation of CIE. Such involvement will reduce the expected resistance for changes, as it will also help in creating user commitments to the CIEs.

7.2.3 Cost–benefit analysis

Cost–benefit analysis is a systematic quantitative method of assessing the desirability of system development projects or policies, when it is important to take a long-term view of future effects and a broad view of the possible side effects prior to implementing the system. Since some organisations see the requirements related to project management, they do not place the cost–benefit analysis in their requirements documents. However, it should be included in the requirements specification. Furthermore, Emam and Madhavji (1995) and Arayici *et al.* (2004) identified the cost–benefit analysis as a factor that affects the success and failure of the RE process. In addition, including cost–benefit analysis in the requirements specification of the CIE development can be a sustainable advantage to convince the construction practitioners for the implementation of the CIE system. Added to this, everyone involved in the cost–benefit analysis needs to understand the process workflow, because it is the baseline for nearly all decisions regarding new IT solutions. Therefore, the process must be thoroughly documented for cost–benefit analysis. In order to assess the cost and to express this as a monetary amount, or time to build, the requirements cost for the development of the system is estimated using

Table 7.3 Measurable deliverables for cost–benefit analysis in requirements specification.

1	Number of input and output flows in the work context
2	Number of business events
3	Number of use cases
4	Number of functional requirements
5	Number of non-functional requirements
6	Number of requirements constraints
7	Number of function points

Table 7.4 Issues considered in cost–benefit analysis.

1	The scope of the cost–benefits analysis
2	The extent to which top management is convinced that the expected benefits are likely to materialise
3	The fit between the available funding profile and the necessary funding profile to implement the CIE system
4	The amount of benefits that are expected to be brought to the organisation by implementing the CIE system
5	The accuracy of cost estimates compared to the accuracy required by the organisation
6	The extent to which the presentation of the cost–benefits analysis follows the accounting procedures of the organisation
7	The appropriateness of the approaches taken to quantify the intangible benefits

one of the estimating methods. However, there is no best method to use for estimating (Robertson and Robertson, 1999). Thus, the estimates should be based on some tangible, countable artefact. For example, a number of measurable deliverables are given in Table 7.3, which can be utilised for cost–benefit analysis.

As a result of requirements costing as well as some investment, benefits should be achieved when the system is implemented. Consequently, the cost is financially justified with the resulting benefits. The more detailed the work is on the requirements, the more accurate will be the deliverables and cost estimate (Robertson and Robertson, 1999). The criteria in Table 7.4 involving cost–benefit analysis are considered for the evaluation of the CIE systems.

7.2.4 The quality of architecture of the CIE systems

Developing system requirements into viable software architecture that satisfies these requirements is an important aspect. This is because

Table 7.5 Issues considered for analysis of the quality of architecture of the CIE systems.

1	The clarity between the (process and data) models and the system objectives
2	The clarity of the business process in the architecture
3	The clarity of links between the (process and data) models and key issues
4	The thoroughness with alternative solutions were explored
5	The clarity between the weaknesses and the strengths of the existing systems and the weaknesses and strengths of the CIE system
6	The adequacy of the diagnosis of the existing system
7	The extent to which key issues have been resolved
8	The extent to which the process and data models conform to the rules of modelling

the level of success in evolving from informal requirements into the formal system structure will reflect the quality of the architecture of the system. The key issues identified in Chapter 5 mainly addressed the requirements modelling and system architecture, which also indicates the thoroughness with which the researchers and engineers approached this aspect. The criteria identified for evaluation under the quality of the architecture group are included in Table 7.5.

These criteria denote the process and product modelling in RE. In fact, process modelling depicts real-world activities of people and their use of the enabling technology to perform work faster, better and cheaper. Furthermore, process modelling also leads to a better understanding of requirements management and better metrics development, which are typically productivity (process time), quality (meets or exceeds customer needs) and cycle time.

In addition, strategic IT implementation should be process led and not technology led (Betts, 1999). This is certainly true; for example, when the process awareness improved in Case Study 3, a better understanding of the requirements and more effective and productive development occurred. Furthermore, previous studies have also shown that technology push is not sufficient to improve the efficiency and effectiveness of the working environments without clear consideration of the business processes and the human issues.

Hence, the clarity of the links between the process model, data model and the system objectives, the clarity of the business process in the architecture and the clarity between process models and key issues of the business are important aspects; these must be taken into account in the RE process to successfully reflect business processes into the system architecture. In order to increase this reflection, it is necessary to employ the correct techniques and methods in the RE process.

Table 7.6 Issues for analysis of the cost effectiveness of requirements engineering process.

1	The cost and effort on the requirements engineering processes
2	The portion of the requirements engineering cost compared to the total project cost
3	The amount of changes in the requirements engineering documentation
4	The amount of deliverables that were not used in design of the CIE System

7.2.5 Cost effectiveness of the requirements engineering process

The last factor for the evaluation of the RE techniques is the cost effectiveness of the RE process. The cost effectiveness analysis can be described as comparing the costs of alternative means for achieving the same stream of benefits.

The cost of the RE activities should be in reasonable proportion to the overall development. This is because very costly requirements activities and techniques may result in the project team omitting some other tasks or duties in the RE and may subsequently result in some ineffectiveness and concerns, with the possibility of costly fixings in the latter stages of the development. Furthermore, adopting expensive RE activities will also increase the product cost of the whole development and will negatively affect the cost–benefit justifications, which will not favour the end users of the product, and may make the end users look at alternative solutions.

To attract the developers to carry out a complete and effective RE process, which will subsequently contribute to the successful demonstration and implementation of the CIE systems, the implementation of an RE technique in the CIE development should involve a reasonable cost. The criteria defined for this aspect are seen in Table 7.6.

7.3 The Evaluation and Assessment in Case Study 3

In order to make the evaluation robust, consistent, relevant and accurate, the evaluation is based on a survey conducted amongst the industrial and academic representatives who were knowledgeable and aware of the developments in Case Study 3. The survey questionnaire covered the 34 criteria, which are clustered in five categories explained in the previous section. Furthermore, these criteria have been evaluated and verified in Chapter 5. Since this set of criteria is designed for measuring the success of the RE processes, it will identify precisely the strengths and weaknesses of the RE practices in Case Study 3.

Respondents are requested to mark each criterion in the questionnaire from 0 to 5 according to their measure of the RE experiences in Case Study 3 against the corresponding criteria. The marks for the criteria are determined as follows:

- 5: Very good
- 4: Good
- 3: It is OK
- 2: Bad
- 1: Very bad
- 0: Not applicable

According to this determination, if any criterion is marked as 3 and above, the RE practice passes that criterion, and mark 3 represents the borderline to pass. However, it does not mean that any criterion marked with 3 or 4 does not need improvement. On the other hand, 1 and 2 refer to failure and they absolutely require enhancement. Lastly, 0 means that the corresponding criteria are irrelevant and cannot be applied to the case studies.

A generalised representative sample of respondents can help appreciate the general validity of the findings about the user requirements performance in Case Studies 3 and 4 and it will serve to understand the unique and generally applicable features. Therefore, a general view of the whole respondents regarding the strengths and weaknesses of user requirements in Case Study 3 will be derived and presented from the survey results. The general view will also form the basis for the interpretation of the survey results. The analysis of the survey is carried out in a quantitative manner to yield numerical values to represent the results clearly and in a simplistic way. As a result, comparisons and interpretations are made easily and clearly.

7.3.1 Plotting the survey data

Once the survey data has been collected, raw data and processed data need to be arranged and presented in such a way that the prior information contained in the data can be interpreted. Raw data is in the completed questionnaires. The processed data represents the overall view of the respondents about the strengths and weaknesses of the RE practices in Case Study 3. It is helpful to produce a diagram or graph of the data. Such plots will help to indicate the level of success, strengths and weaknesses in a quantitative manner.

Presentation of the raw data provides the greatest detail. However, even if it is ordered in some way, the data may not be easy to interpret.

$$X = 1/n\ \Sigma x_j$$

Figure 7.1 Formula for arithmetic mean.

In presenting data, it is common for detail to be sacrificed, so that intelligibility is improved by the use of tables and diagrams. Therefore, it is helpful to represent the raw data in the appendices, which is also valuable for further studies, tests, etc, and for checking and verification of the instant analysis and results.

On the other hand, the processed data that is the arithmetical mean value of the marks given for each criterion by the respondents is calculated using the formula seen in Figure 7.1 and represented in Table 7.7 in the next section. In the formula above X refers to the average mark value for the corresponding criterion, n refers to the number of respondents, j is the increment from 1 to n and x_j refers to the mark value given by the corresponding respondent.

Secondly, the mean marks for each criterion are represented in a bar diagram, which has the particular property that the vertical length of each rectangle in the diagram represents the numerical value of the mark given for the corresponding criteria, which makes the interpretation of the survey results easy. In addition, the respondents were not asked to rank the criteria, although each criterion has different weighting. This is justified as follows:

- These criteria have already been ranked and prioritised through a broader research study.
- The order of the clusters and the criteria in each cluster are listed from top to bottom according to their rank and priority in both Chapter 5 and the survey questionnaire.
- Asking the respondents for ranking the criteria would bring about subjective and misleading results because they would only rank the criteria based on their experiments. Furthermore, most of the respondents who were involved in the Case Study 3 project are not very familiar with the RE profession and system development aspects. As a result, they are not the target audience to rank the criteria.
- Ranking will be one of the aspects in the interpretation of the survey results later in this chapter.

7.4 Survey Results and Evaluation

The survey questionnaire data was collected from people across Europe, which was comprised of contractors, engineers, architects,

Table 7.7 The survey questionnaire filled in with the average marks.

		5	4	3	2	1	0
The quality of requirements engineering process in Case Study 3							
1	**Match of the CIE system with construction companies**	5	4	3	2	1	0
C1	The capability of companies to make changes for uptake			X			
C2	The match between CIE system and strategic orientation of the companies		X				
C3	Senior management support for changes for uptake			X			
C4	The willingness of companies to make changes for uptake			X			
C5	The match between CIE system architecture and corporate structure of organisations in the construction projects			X			
C6	The match between CIE system and technical orientation of the companies			X			
2	**User satisfaction and commitment**	5	4	3	2	1	0
C7	Level of understanding of end users about what CIE system can/should do and will not do		X				
C8	The willingness of users to defend CIE system in front of executive management			X			
C9	The extent of user consensus on the CIE system			X			
C10	Awareness of users about the business changes for uptake			X			
C11	The relationship between stakeholders and requirements engineering team			X			
C12	The users' reaction to the cost estimate					X	
C13	The fit between the architecture and the way the users work			X			
C14	Whether the users have approved all the documentation				X		
C15	The level of match between functionality of alpha release and user expectations			X			
The quality of the CIE system							
3	**The quality of cost–benefits analysis**	5	4	3	2	1	0
C16	The completeness of scope of the cost–benefits analysis					X	
C17	The level of convincing of executive management that expected benefits can be materialised				X		
C18	Fit between the available funding and the funding required for uptake					X	
C19	The amount of benefits to the construction stakeholders by uptake			X			

(continued)

Table 7.7 (*Continued*)

C20	The accuracy of the cost estimates relative to the accuracy required by organisations					X	
C21	To what extent the cost–benefits analysis follows the accounting procedures of organisations					X	
C22	Appropriateness of approaches taken to quantify intangible benefits when the system is implemented					X	
4	**The quality of the architecture of the CIE systems**	5	4	3	2	1	0
C23	The clarity between the (process and data) models and the system objectives		X				
C24	The clarity of the business process in the architecture			X			
C25	The clarity between the process and data models and the key issues			X			
C26	The thoroughness with which alternative solutions were explored			X			
C27	The clarity of the links between the weaknesses and strengths of the existing technologies and the weaknesses and strengths of the CIE systems			X			
C28	The adequacy of the diagnosis of the existing system				X		
C29	The extent to which key issues have been resolved			X			
C30	The extent to which the process and data models conform to the rules of modelling				X		
Cost effectiveness of the RE process							
5	**Cost effectiveness of the RE process**	5	4	3	2	1	0
C31	Cost and effort on the requirements engineering process				X		
C32	Proportion of the cost of requirements engineering compared to the estimated total system development cost				X		
C33	Amount of changes in the requirements engineering documentation				X		
C34	Amount of deliverables that were not used in the system designing of CIE				X		

developers, academics and researchers. They were all well informed and knowledgeable about the CIE development in Case Study 3. Table 7.7 shows the average marks given for each criterion.

It is observed that marks given by the user group of the project are relatively lower than the marks given by the developers of the project, which may imply that the technical people are more optimistic than the end users. In other words, the level of awareness and perception of

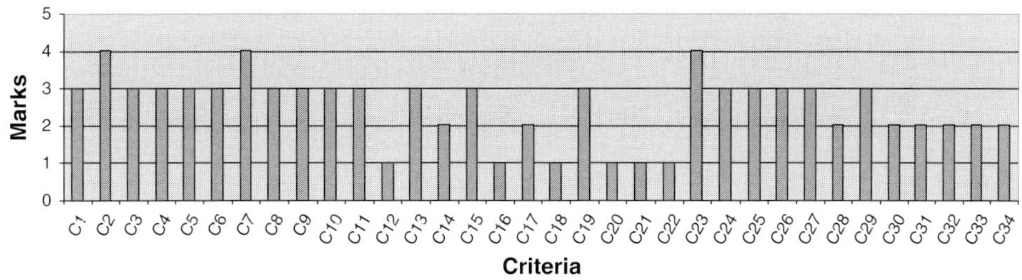

Figure 7.2 Survey results of the requirements engineering practice in Case Study 3.

the end users and that of the technical people are not the same. This can be attributed to the lack of communication and lack of shared understanding in the project. This issue will be elaborated on later in this section. The bar graph in Figure 7.2 represents the average marks given for each criterion.

The graph above shows that the RE approaches adopted in Case Study 3 are not completely successful. The majority of the values are on the borderline while the rest are below the borderline; there are three criteria that are marked as 4. These criteria are C2, C7 and C23, which are the only strengths of the RE process. While C2 is classified in the match of the CIE systems with construction industry group, C7 and C23 are classified in the user satisfaction and commitment and the quality of the architecture of the CIE systems groups, respectively.

It is an interesting result because all three criteria are process related. Because half way through the project it was the storyboard process that led to the development until the end, CIE developments were process led in Case Study 3. Furthermore, Sheath *et al.* (1996), Auoad *et al.* (1998), Betts (1999) and Aouad and Wafai (2002) highlighted the process-led development, which is strategically important in future CIE developments. On the other hand, while 17 criteria are marked as 3, 8 criteria are marked as 2 and 6 criteria are marked as 1. In other words, only 3 criteria pass the borderline, 17 of them are on the borderline and 14 of them are below the borderline. Under these circumstances, the implementation of the RE process is inclined to be successful. However, there are still many criteria in terms of which the RE approaches in Case Study 3 for RE process failed. The graph in Figure 7.3 shows the distribution of marks for the criteria.

Mark 3 indicates borderline success; however, it does not mean that the criteria assigned at the borderline do not need enhancement, but that they need improvement for persistent success. If the criteria that were appointed to the borderline were put aside, the result would

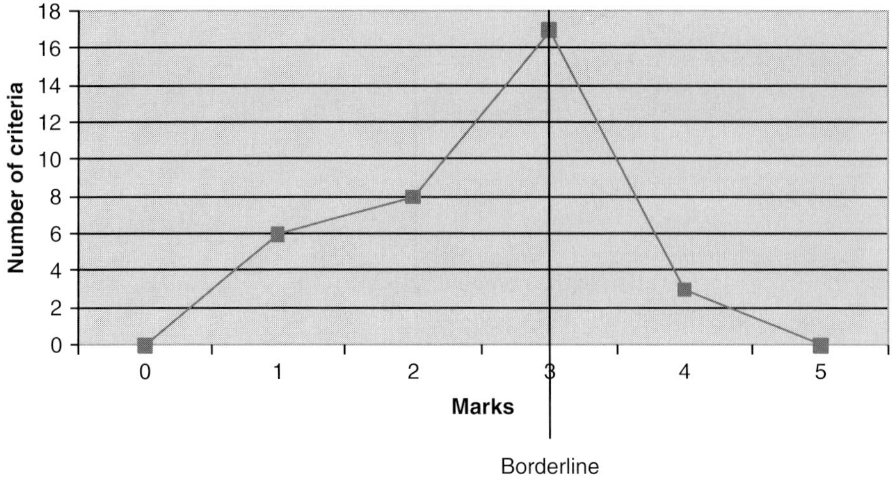

Figure 7.3 The distribution of marks over the criteria.

be seen as a failure. This is because only 3 criteria are above the borderline, whereas 14 criteria are below the borderline. As a result, the RE practice in Case Study 3 can be asserted as an ad hoc implementation, although the techniques and methods (contextual design (CD) technique, use case-driven analysis and incremental prototyping as an agile process) are already proven and validated as state-of-the-art RE techniques. To enhance the performance of the RE activities, it is certainly required to increase the success rate to be above 3, which subsequently results in more successful CIE systems and contribution to its implementation. In the following sections the analysis is carried on to address the specific key issues, strengths and weaknesses.

7.4.1 Comparing the views of the technical and user respondents

The questionnaire was sent to the people who were knowledgeable about the CIE developments in Case Study 3. Some were technical people from Construct IT background, while some others were construction practitioners. Tables 7.8 and 7.9 show the average marks given by the practitioners and technical people respectively.

In order to ease the comparison, the average marks are represented in graphs in Figures 7.4 and 7.5 based on Tables 7.8 and 7.9 respectively.

The graphs in Figures 7.4 and 7.5 reflect the understanding and perceptions of the user team and the technical team of the project. From the graphs, it can be said that there is not much shared understanding between the two groups. This is attributed to the lack

Table 7.8 The average marks given by the construction practitioners.

The quality of the requirements engineering process in Case Study 3							
1	**Match of the CIE system with the construction industry**	5	4	3	2	1	0
C1	The capability of companies to make changes for uptake			X			
C2	The match between CIE system and strategic orientation of the companies			X			
C3	Senior management support for changes for uptake			X			
C4	The willingness of companies to make changes for uptake			X			
C5	The match between CIE system architecture and corporate structure of organisations in the construction projects			X			
C6	The match between CIE system and the technical orientation of the companies			X			
2	**User satisfaction and commitment**	5	4	3	2	1	0
C7	Level of understanding of end users about what CIE system can/should do and will not do		X				
C8	The willingness of users to defend the CIE system in front of executive management		X				
C9	The extent of user consensus on the CIE system			X			
C10	Users' awareness about the business changes that are required for the CIE adoption			X			
C11	The relationship between the users and the requirements engineering staff			X			
C12	The users' reaction to the cost estimate					X	
C13	The fit between the architecture and the way the users work			X			
C14	Whether the users have approved all the documentation						X
C15	The degree of match between the functionality of the 1st release of the system and user expectations			X			
The quality of the CIE system							
3	**The quality of cost–benefits analysis**	5	4	3	2	1	0
C16	The scope of the cost–benefits analysis					X	
C17	The extent to which top management is convinced that the expected benefits will be materialised				X		
C18	Fit between the available funding and the funding required for uptake				X		
C19	The amount of benefits to the construction stakeholders by uptake				X		

(continued)

Table 7.8 (*Continued*)

C20	The accuracy of the cost estimates compared to the accuracy required by the organisation						X
C21	To what extent the cost–benefits analysis follows the accounting procedures of organisations						X
C22	Appropriateness of approaches taken to quantify intangible benefits when the system is implemented					X	
4	**The quality of the architecture of the CIE system**	5	4	3	2	1	0
C23	The clarity of the links between the (process and data) models and the system objectives		X				
C24	The clarity of the business process in the architecture			X			
C25	The clarity of links between the (process and data) models and the key issues				X		
C26	Attention given to the alternative solutions for CIE system to be developed			X			
C27	Clarity of links between weaknesses and strengths of Information and Communication Technology (ICT) systems in use and weaknesses and strengths of CIE system to be developed			X			
C28	The adequacy of the diagnosis of the existing system					X	
C29	The extent to which key issues have been resolved				X		
C30	The level of compliance of process and data models to the rules of modelling					X	
Cost effectiveness of the RE process							
5	**Cost effectiveness of the RE process**	5	4	3	2	1	0
C31	Cost and effort on the requirements engineering process					X	
C32	The portion of the requirements engineering cost compared to the total project cost					X	
C33	The amount of changes in the requirements engineering documentation					X	
C34	Amount of deliverables that were not used in system designing of CIE					X	

of communication and interaction between the users and technical participants. Although there is a rough level of shared understanding in the criteria from C1 to C11, there are slight differentiations. For example, the developers assigned C3 to failure, the users marked it as 3 but the developers marked the rest of the criteria from C1 to C11, either as the same as the users or higher than the users. For example, many criteria from C1 to C11 are marked with 3 by both the users

Table 7.9 The average marks given by the technical people from the construct IT background.

The quality of requirements engineering process in Case Study 3							
1	**Match of the CIE system with construction companies**	5	4	3	2	1	0
C1	The capability of companies to make changes for uptake			X			
C2	The match between CIE system and the strategic orientation of the companies		X				
C3	Senior management support for changes for uptake				X		
C4	The willingness of the organisations to make the necessary changes to implement the CIE system			X			
C5	The match between CIE system architecture and corporate structure of organisations in the construction projects			X			
C6	The match between CIE system and technical orientation of the companies		X				
2	**User satisfaction and commitment**	5	4	3	2	1	0
C7	Level of understanding of end users about what CIE system can/should do and will not do		X				
C8	The willingness of users to defend CIE system in front of executive management			X			
C9	User consensus on CIE system			X			
C10	Awareness of users about the business changes for uptake			X			
C11	The relationship between the stakeholders and requirements engineering team			X			
C12	The users' reaction to the cost estimate				X		
C13	The fit between the architecture and the way the users work			X			
C14	Whether the users have approved all the documentation		X				
C15	The level of match between functionality of alpha release and user expectations			X			
The quality of the CIE system							
3	**The quality of cost–benefits analysis**	5	4	3	2	1	0
C16	Completeness of the cost–benefits analysis				X		
C17	The level of convincing of executive management that expected benefits can be materialised				X		
C18	Fit between the available funding and the funding required for uptake					X	
C19	The amount of benefits to the construction stakeholders by uptake			X			

(*continued*)

Table 7.9 (*Continued*)

C20	The accuracy of the cost estimates compared to the accuracy required by the organisation				X		
C21	To what extent the cost–benefits analysis follows the accounting procedures of organisations					X	
C22	Appropriateness of approaches taken to quantify intangible benefits when the system is implemented				X		
4	**The quality of the architecture of the CIE system**	5	4	3	2	1	0
C23	The clarity of the links between the (process and data) models and the system objectives		X				
C24	The clarity of the business process in the architecture			X			
C25	The clarity of links between the (process and data) models and the key issues			X			
C26	The thoroughness with which alternative solutions were explored			X			
C27	Clarity of links between weaknesses and strengths of ICT systems in use and weaknesses and strengths of CIE system to be developed		X				
C28	The adequacy of the diagnosis of the existing system			X			
C29	The extent to which key issues have been resolved		X				
C30	The level of compliance of process and data models to the rules of modelling			X			
Cost effectiveness of the RE process							
5	**Cost effectiveness of the RE process**	5	4	3	2	1	0
C31	Cost and effort on the requirements engineering process			X			
C32	Proportion of the cost of requirements engineering compared to the estimated total system development cost			X			
C33	Amount of changes in the requirements engineering documentation			X			
C34	Amount of deliverables that were not used in system designing of CIE			X			

and developers. Furthermore, while developers assigned C2, C6 and C7 with the mark 4, the users assigned C7 and C8 with the mark 4.

From C12 to C22, the level of understanding of users and developers fluctuates. For example, C13 is marked as 3 by both the users and developers; C14 is marked as 4 by the developers whereas it is marked as 0 by the users. This reflects the fluctuation and deep differentiation in the shared understanding. Furthermore, the criteria C20 and C21

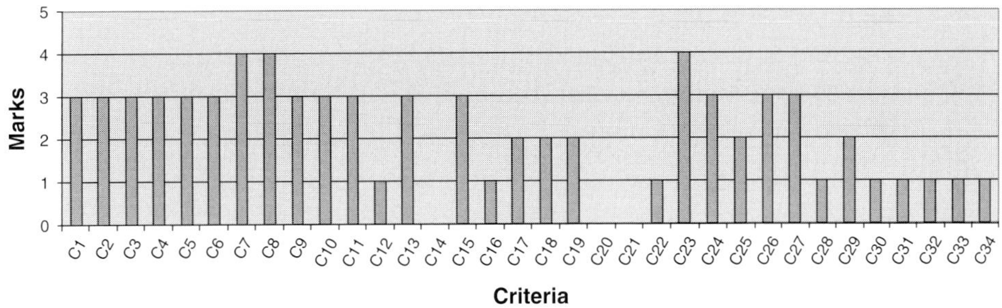

Figure 7.4 The average marks given by the construction practitioners only.

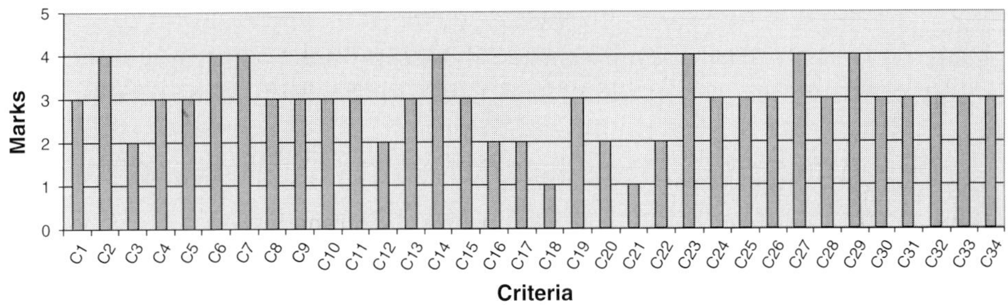

Figure 7.5 The average marks given by the technical people from Construct IT background.

are marked as 0 by the users, while the same criteria are marked as 2 and 1 respectively by the developers. This can also be interpreted as a differentiation in the shared understanding. In addition, the shared understanding of both the users and the developers represents similarities from C22 to C27. According to the users it is a complete failure from C28 to C34, while it is relatively bearable according to the developers. Again, this represents the differentiation in the shared understanding between the users and developers because of a lack of communication, interaction and collaboration in the process.

Another interesting result is the differences in the approaches of the two groups; the technical people are more optimistic and the RE process is relatively successful compared to the users' perspectives, which are less optimistic. Furthermore, according to the users' views, the RE process, which was conducted in an ad hoc manner, is not successful. Subsequently, the users could not be aware of the development process and some aspects and key issues could not be addressed properly, and hence remained vague for the users.

Overall marking by the users, according to the formula in Figure 7.1 is 2, whereas it is 3 according to the developers. It means that the RE practices in Case Study 3 were not successful and needed enhancement to be successful, according to the end user respondents. On the other hand, according to the developers, they were on the borderline and assumed as successful but still needed improvements.

In the following section, the analysis is continued according to the groups defined in Chapter 5.

7.4.2 Fit of the CIE system in Case Study 3 with the construction industry

This groups is comprised of the criteria from C1 to C6. The marks given in this category are slightly better than the other groups. For example, while C2 is marked as 4, the rest in this group are marked as 3 (see Figure 7.2). The overall success rate or average mark value of this group is calculated as 3 using the formula of arithmetic mean shown in Figure 7.1. Therefore, it can be accepted as being relatively successful compared to the other groups. Below are the interpretations and observations about the criteria in this group.

The main strength of the RE practices in Case Study 3 is being a process push development, which resulted in being marked as 4 for the criteria (C2), which is the fit between the CIE systems and the strategic orientation of the construction organisations.

A respondent from the developer group conveyed that the acoustic and thermal simulation modules were totally implemented and integrated in a collaborative virtual reality (VR) platform in France. Furthermore, case studies contributed to the strategic orientation of VR development and business modelling. Another respondent from the user group highlighted that organisations are not ready for the CIE systems as a whole system for the time being. Employing the CIE systems will require organisational changes and a redefinition of responsibilities. On the other hand, few companies are strategically working on a move towards partnering or similar collaborative paths.

Owing to the uncontrollable external parameters, such as the traditionally fragmented nature, a different level of knowledge, perspectives and process issues, etc., the fit between CIE developments in both Case Study 3 and the strategic and technical orientations of the organisations is weak at the moment.

The CIE system deals with the core business of architects, 4D simulation and site monitoring for contractors and simulation, core calibration and calculation applications for engineers. Although none of these practitioners need to have the whole CIE system, they can still collaborate using the only relevant applications of the CIE system through the communication framework even if they are situated in different companies in different locations. The component-based architecture of CIE systems in Case Study 3 eases the fit between the system and the construction organisations.

However, the CIE system also varies in itself in terms of this fit. For example, in Case Study 3 while the engineering applications (simulation applications) are marked as 4, contractor (4D simulation and site planning and analysis) and architectural applications (Client Briefing) are marked 3 and 2 respectively. That is to say, in Case Study 3 while engineering applications provide a good fit with the relevant construction organisations, contracting applications provide a lower fit than engineering applications, but architectural applications cannot provide the sufficient level of fit. That is also attributable to imbalance of the focus and clarity in the RE process.

7.4.3 User satisfaction and commitment

This group comprises the criteria from C7 to C15. The marks given in this category are relatively better except for C12 and C14, which are marked 1 and 2 respectively. C7 is marked as 4 and the remaining criteria are marked as 3 (see the Figure 7.2). The overall success rate or the average mark in this category is calculated as 3 using the formula of arithmetic mean shown in Figure 7.1. However, the mark is rounded upwards from 2.78. Therefore, the RE practices in Case Study 3 were better in the previous category than in this category. As a result, it can be accepted to be relatively successful compared to the other categories.

C12, which is about cost estimate, was not addressed in detail in Case Study 3. However, the cost estimate of the development and deployment should be taken into consideration in the RE processes, which can produce encouraging results to convince the end users to employ the CIE systems in the future. On the other hand, C14, which is about approving all documentation by the users, was not clearly followed in the case studies. This is because the users were not fully aware of the development process. Furthermore, they did not really know what to expect next until they experienced the initial prototypes in the project. Therefore, the documentation was not really under the control of the users until the testing processes.

Some respondents commented that there was still a need for further improvement with regard to user satisfaction and commitment. The users realised that they had to make significant improvements too. They were only aware of this towards the end of the project. Earlier in the project their knowledge was not sufficiently enough, nor was the specification of what the CIE systems might do clear. In addition, the relationship between the users and RE staff was good but at the beginning of the project the users and technical staff had separate workshops during three monthly meetings. In hindsight, this was not a good idea.

The first release was pretty simplistic and the users had hoped for more. Furthermore, conducting collaborative sessions between the architectural and engineering applications (acoustic, thermal, lighting) helped to discover the need for stronger communication. In addition, the communication layer for virtual communication and collaboration had no useable interface for virtual collaborative environments. Too much effort was put into the software applications of the CIE systems in Case Study 3 but less effort was spent on the communication and integration tools.

With regard to the fit between the architecture and the way the users work, some user respondents commented that some parts of the CIE system in Case Study 3 fit well with the way users work, but some other parts did not fit well with the way users work in the industry. For example, in reality, simultaneous collaborative work is not possible on the same building model, which implies the use of sophisticated IT technology to increase the productivity and accuracy. These features of the CIE system in Case Study 3, for example, have the potential to be used in future.

7.4.4 Quality of cost–benefit analysis

This group comprises the criteria from C16 to C22. The marks given in this category are the worst ones. C16, C18, C20, C21 and C22 are marked as 1, C17 is marked as 2 and C19 is marked as 3 (see the Figure 7.2). The overall success rate or the average mark value of this group is calculated as 1 using the formula of arithmetic mean shown in Figure 7.1. Therefore, it is a complete failure of the RE practices in Case Study 3 for this category.

Because they were research projects the cost–benefit analysis was not taken into account in the RE practices of these case studies at all. Therefore, the respondents struggled a lot to judge and decide the correct value of the marks and provide some comments in this category. Furthermore, asking the respondents to mark the RE practices in the case studies in terms of cost–benefit analysis made some of them hesitant and suspicious and some felt a lack of confidence and thought that they should have taken into consideration the cost–benefit analysis, and as a result, all these views also refer to a lack of awareness and communication because of the ad hoc users' involvement in the RE process.

If this analysis was considered in the RE process, some figures for return on investment (ROI) could have been generated, which would have at least encouraged the construction practitioners to employ and implement the system in the future, or at least minimise the level of resistance against the implementation of such systems.

7.4.5 The quality of the architecture of the CIE system

This group consists of the criteria from C23 to C30. The marks given in this category can be classified as the second best marks because C23 is marked as 4 and C24, C25, C26, C27 and C29 are marked as 3. However, C28 and C30 are marked as 2 (see the Figure 7.2). The overall success rate or average mark for this group is calculated as 3 using the formula of arithmetic mean shown in Figure 7.1. However, the mark is rounded upwards from 2.88. Therefore, it can be accepted as relatively second in success compared to the other groups.

C28, which is about the adequacy of the diagnosis of the existing systems, was not clearly addressed in the project although the users were well aware of the existing systems. Existing commercial systems would have been investigated and their adequacy could have been analysed in order to embed these systems into the CIE system platform of the Case Study 3 projects, which would have led to commercial off-the-shelf (COTS)-based RE. COTS-based RE is cost effective but limits the flexibility of system functionalities and architecture according to the capabilities and features of the commercially available tools. Furthermore, it could enable an alliance between the vendors of the existing systems and the CIE community.

C30, which is about the conformity of the process and data models to the rules of modelling, was not understood well by the respondents, especially the end users. That is why they marked this criterion as low. Furthermore, process and data modelling reminds some respondents only of Industry Foundation Classes (IFC) modelling. However, process and product modelling was meant to be the business process and the system design.

Some respondents did not understand the criteria at all. Actually in this criterion, it was meant that a process model includes processes of the software system (functions) as well as the manual work activities of the business process and data modelling, for example, how the business process is translated into a data model that also reflects the architecture of the system. In short, the criterion was actually addressing the rules of techniques used in the RE approaches in Case Study 3 including CD, use case development, storyboarding and testing, etc. Once more, the lack of awareness and understanding of the users about the RE techniques are pointed out.

Some respondents indicated that the overall picture became clear for most of the participants after 2 years in the project and there was still some confusion until the final stage of testing. In addition, according to some respondents there was no key issue clearly identified by the end users and the requirements models were developed to be in line with users' potential ways of working with the new system.

On the other hand, some other respondents conveyed that the key issues partly changed during the project, from workspace/collaboration issues to focus on underlying building product models and infrastructures to support activities in the distributed workspaces.

In terms of the clarity of the business process in the architecture of the system, the process model was based on user requirements mediated by the system capabilities and system objectives. Hence, the system was based on the business process.

7.4.6 Cost effectiveness of the requirements engineering process

This group embraces the criteria from C31 to C34. The marks given in this category can be classified as the second worst set of marks because all criteria in this category are marked as 2 (see Figure 7.2). The overall success rate or average mark for this group is calculated as being 2 using the formula of arithmetic mean shown in Figure 7.1. Therefore, it is a second complete failure of the RE practices in Case Study 3 in this category.

Cost effectiveness was not a priority in the Case Study 3 project, and was not taken into consideration in the requirements activities. Although every user group in the project was using the budget provided for them, not enough study or attention was given to the effectiveness of the requirements activities they conducted.

Although sometimes the requirements documentation was changed to optimise and compromise the requirements, due to different reasons, some users were not aware of these changes. Too many changes are not good in terms of cost, but if there are few changes, they are good in terms of saving resources. Also, it implies that requirements are expressed clearly in the documentation and that users know what they want and what they can have with the available technologies, and developers know what users want and what they can develop.

In short, it means that there is a consistent communication and collaboration between the users and developers and inevitable changes are managed well in requirements management. However, the communication and collaboration issues were the issues in the case study projects, so requirements documents kept changing and there was no opportunity for the users to approve the documentation every time.

7.4.7 Summary of the analysis

As a high-level criticism to the implementation of the RE approaches in Case Study 3, there was a lack of communication and interaction between the parties. There were some strategic mistakes. For example, at the early stages of the project, workshops were conducted by the developers and end users separately, which led to the lack of shared understanding.

The end users lacked an awareness about the user requirements activities, and at the earlier stages of the project the RE process was not well defined and structured. The initial techniques, which were the CD and use case-driven analysis, were conducted in isolation and also in the wrong order. First the use case-driven requirements elicitation and analysis was conducted, and following that the users and developers realised its inadequacy. The project consortium decided to employ additional techniques such as CD and storyboarding. However, the use case-driven analysis should have been done after the CD technique and storyboarding. Consequently, initial requirements activities were ad hoc activities and requirements documentation and specification were not sufficient and well structured. The specific results of the analysis are listed below:

- Although communications and networking will be a major topic in the near future, the emphasis was less for the communication and networking aspects than the specific applications in the RE practices in Case Study 3.
- There was no solid contribution to the vision definition in the RE in Case Study 3, which was simply based on the previous research studies, and the early consecutive requirements activities did not address the issues in the vision definition. The only contribution from Case Study 3 is the workflow definitions in the CD approach. On the other hand, through workshops, a strong vision definition and development was achieved in Case Study 4.
- One of the important results from Case Study 3 in particular is to emphasise the need for parallel integrated efforts on user environment and system design and the greater emphasis on early user needs and requirements capture.

- At the early stages of the projects, implementation of the system was not a concern at all.
- Although it was a research project, it is a prototype of the concept demonstrating BIM that is eventually dominating the construction industry. Therefore, the issue of the fit between the system and the industry should have been taken into account in the very early stages of the definition in the requirements process, as opposed to the latter stages of the project.
- Discussing the issues below with the end users and the developers could have led to a shared vision amongst the participants, and the users could have better conceived the idea of CIE even before testing the prototypes. Furthermore, it could have increased the interaction between the parties in the following areas:
  - The ability of the organisations to make the necessary changes to implement the system
  - The fit between the CIE systems and the technical and strategic orientation of the organisations
  - The fit between the CIE systems architecture and the corporate architecture
  - The willingness of organisations to make the necessary changes to implement the system
- With regard to the fit between the system and the industry, the RE practice was becoming relatively the most successful. This is attributed to the techniques used half way through the project, such as storyboarding and incremental prototyping, with the end users tests as an agile development process that increased the process awareness and entailed the users' involvement intensively.
- In Case Study 3 in particular, since halfway through the project it was the storyboard process that led to the development until the end, the case study was a process-led development, which is the main strength of the RE practice.
- The category of user satisfaction and commitment is the least successful category in the quantitative analysis. Owing to the lack of completeness of the requirements documentation, the users were not sure in many respects until the final stage.
- Owing to a lack of traceability for process and product modelling and a lack of proper structuring and formalisation of the amount of knowledge captured by means of traceability, not many people were sure what the new system did and did not do until the final stage.
- In many respects, an insufficient capability of handling the complexities through the development process led to the users involved in

the project not being able to pursue the development properly, until the storyboard enhancement and users testing.

- Requirements management was initially conducted with a lack of process and product awareness.
- Cost-related aspects in the requirements process were not taken into account.
- The shared understanding between the stakeholders was not evolved until halfway through the project.
- Since many users were still learning about the RE and the CIE concept, the capability to resolve the inconsistencies in requirements was low and insufficient.
- Because the user requirements activities were conducted in an ad hoc manner until halfway through the project, continuous traceability did not properly occur until the storyboard enhancement and user tests were developed as an agile process. Furthermore, this ad hoc manner caused difficulties in enhancing the storyboard and preparing the testing materials. As a result, user consensus on the system, user approval for the documentation, the match between the functionality of the first early release of the system and the user expectations were lower than expected.
- The RE practices in Case Study 3 failed in the quality of cost–benefit analysis because it was not strategically taken into account in the RE activities.
- Cost–benefit analysis did not take place and the constraints and cost requirements were not analysed thoroughly in Case Study 3. It can be justified by the fact that it was a research project; however, a high-level cost–benefit analysis such as an ROI could have convinced the stakeholders to implement the system. Thus, the cost–benefit analysis is a necessity in future for commercially based CIE systems development.
- The quality of the architecture of the system is the second successful category relative to the other categories. This can be attributed to the techniques used halfway through the project that addresses the architecture of the system.
- Testing the requirements through storyboarding against the system enabled relatively better, more consistent and accurate requirements specification.
- The clarity of links between the process model, data model, the system objectives and the clarity of the business process in the system architecture became relatively clearer and improved.
- There were no key issues identified. Although many aspects such as collaboration, distribution, integration, common data exchange, etc.

were stressed, they were not clearly defined as key issues. However, they were apparently the key aspects of Case Study 3; the clarity of links between the process and data models and key issues were improved later in the project but not fully resolved.

- There was no benchmarking with the existing systems or software in use in the construction industry in the RE practices. However, it may have enabled the identification of advantages and disadvantages between the existing systems in use and the case studies' RE practices, and possible overlaps between them could have been benefited.
- Cost effectiveness of the RE processes is the least influential factor. However, it is due to the second factor that the RE process of Case Study 3 failed. For a research project, measuring the cost effectiveness of the RE processes is not a priority because the required budget for requirements activities is already defined in advance. However, doing the cost effectiveness analysis of the utilised RE techniques and methods in a research project can help increase the productivity of the techniques with the same budget.
- On the other hand, for commercially based development, keeping the cost reasonable for the RE activities is important. From that perspective, cost effectiveness is relatively more important for commercially based developments than research-based developments.

References

Aouad, G., Hinks, J., Cooper, R., Sheath, M.D., Kagioglou, M., Sexton, M., (1998), "An IT Map for a Generic Design and Construction Process Protocol", Journal of Construction Procurement, 4(2):132–151.

Aouad, G., Wafai, M.H., (2002), "Implementation of Information Technology in the Construction Industry: The Conceptual Methodology Approach", 2nd International Postgraduate Research Conference In the Built and Human Environment, University of Salford, Greater Manchester, United Kingdom.

Arayici, Y., (2004), "Requirements Engineering in Innovative Systems Development for the Construction Industry", PhD thesis, SCPM, University of Salford, Greater Manchester, United Kingdom.

Betts, M., (1999), "Strategic Management of IT", Blackwell Science, ISBN 0 632 04026 2.

Damain, E.D., (2003), "A research methodology in the study of requirements negotiations in geographically distributed software teams", 11th IEEE International Requirements Engineering Conference (RE'03), Monterey, California.

Emam, K., Madhavji, N., (1995), "Measuring the Success of Requirements Engineering Processes", Second IEEE International Symposium on

Requirements Engineering, York, United Kingdom, IEEE Computer Society Press, pp. 204–214.

Herlea, D.E., (1996), "User Involvement in Requirements Engineering Process", Tenth Knowledge Acquisition for Knowledge-Based Systems Workshop (KAW96), Banff, Canada.

Paulson, B.C., (1995), "Computer Applications in Construction", Civil Engineering Series, McGraw-Hill International Editions.

Robertson, S., Robertson, J., (1999), "Mastering Requirements Engineering Process", Boston, Massachusetts, ACM Press, Addison Wesley, ISBN 0201360462.

Sheath, D.M., Woolley, H., Cooper, R., Hinks, J., Aouad, G., (1996), "A Process for Change–The Development of Generic Design and Construction Process Protocol for the UK Construction Industry", Proceedings of CIT Conference '96, Sydney.

8

Mastering the Requirements Engineering Practices

8.1 Introduction

Despite a constant improvement in the implementation of the user requirements techniques in computer integrated environment (CIE) developments, the level of guidance for user-driven and also industry-driven CIE development is still very low. More effective guidance needs to be determined to deal with a large variety of situations. In this chapter, by taking into consideration the experiences from Case Studies 3 and 4 as well as the experiences from Case Study 2, the requirements engineering (RE) practices will be improved to act as an RE process framework for the CIE developments and implementation.

There are five different causes that result in failure or under-delivery in CIE developments. These are (i) the RE process itself, (ii) human communication within the RE process, (iii) knowledge development, (iv) documentation of requirements and (v) management. For example, the evaluation results of RE practices in Case Study 3 in Chapter 7 can be easily clustered under these five causes. Furthermore, these five causes are also greatly related to the key issues and criteria elaborated on in Chapter 5. By remedying these five causes, a generic framework can be determined.

8.2 Project Start-off

The start-off meetings are joint application development (JAD) meetings of the project steering committee, which should work together until they achieve the start-off objectives. The steering committee, which is involved in the start-off meeting, comprises the client, the main users, lead developers, technical and business experts and other people who are key to the success of the project. They should have brainstorming sessions to identify all the stakeholders, who have requirements for the CIE development. If all the stakeholders are not defined, it is likely that some of the requirements are missed. A preliminary estimate of the cost is also produced in the start-off meeting, as well as an early risk assessment that the project is likely to face.

8.2.1 Setting the scope

The scope of the project is the shared scope of the stakeholders identified for the CIE development. The fragmented duties of practitioners in a multidisciplinary work environment are aimed at integration by means of the CIE systems through a lean process. If the scope is clearly understood by the developers, they can then build the CIE systems successfully. Setting the scope also means defining the boundaries of the CIE development. It sets constraints on what should be inside and what can be safely left outside the development. The project team must ensure that everything within the scope is sufficient to meet the objectives set for the solution; similarly the budget available and the time allowed are also related to the scope.

8.2.2 First version of the requirements specification deliverable

The first version of the requirements specification deliverable is produced at the end of the start-off phase and distributed to all the related parties. It explains the outputs from the start-off phase and lays the foundation for the project and covers the following:

- Goals of the project – short, measurable statements of what the CIE system must achieve for business
- Stakeholders who are interested in the CIE system
- Constraints – the limitations that the project is likely to face
- Names – terminology to be used in the project
- Scope of the work – the boundaries of the product and the project
- Estimated cost – the amount of effort or money that must be expended to bring the product to fruition
- Risks – a short risk analysis to reveal the main risks faced by the project and cost–benefit analysis for return on investment (ROI)

8.3 Requirements Elicitation

The requirements engineers initiate the requirements elicitation and should collaborate with the users and stakeholders to gather the requirements. It uses the outputs of the project launch as a starting point for collecting requirements from the users. The requirements elicitation activity is multi-faceted. The requirements engineer should understand the work of the practitioners and determine a new work redesign that the practitioners wish to use in the future that will utilise the CIE systems in their workplace. Elicitation activity is both a study of the work and a means of improving the ways in which the practitioners manage their duties.

The contextual inquiry from Case Study 3 is the main requirements elicitation technique. The requirements analyst conducts one-on-one interviews with the end users and observes them as apprentices whilst they are working. However, this does have drawbacks for the CIE system developments. Multiple-use locations of the construction industry can increase the costs and the amount of research time. This can be a disadvantage in terms of the cost effectiveness of the RE process. Furthermore, Christianson (2001) also raised the same issue that the distribution of the end users in the design team, which demonstrates the multi-disciplinary stakeholders, across Europe obstructed a full contextual inquiry process. However, it would be ideal if the contextual inquiry was utilised as long as the project scope and resources available allowed it.

Considering the worst case scenario in a CIE development project, instead of the contextual inquiry, the requirements elicitation should include techniques such as interviews, workshops and work modelling, as validated by the positive experiences in Case Studies 2, 3 and 4. The following subsections elaborate on these techniques for requirements elicitation.

8.3.1 Interview the users

Interviewing the users is the traditional approach to requirements gathering. Interviews enable the analyst to get feedback of his/her understanding and build work models while interviewing the users. The interview technique was used in Case Studies 2 and 3.

8.3.2 Workshops

The intention of the workshop is to transfer knowledge of the business event from its experts to the requirements engineer, where any interested parties and scenarios and relevant work models are constructed to show the actions necessary. This type of workshops were of significant benefit in Case Study 4. The deliverables from the workshops are as follows:

- Purpose of the business event
- Scenarios of the work done to respond to the business event
- 'What if' scenarios when the event happens
- The business rules that apply to that part of the work
- The likely users of the system to be built for this business event

8.3.3 Brainstorming

The CIE system should include new and better ideas for improving the work. Brainstorming is a method for generating new ideas. In a brainstorming session the stakeholders discuss and generate new ideas for the product. Some basic advice for brainstorming sessions are as follows:

- Participants should come from a variety of disciplines. This mixture of background brings many more creative ideas.
- Instead of judgement, evaluation and criticism, the requirements engineer should only record requirements as they are generated, which enables the fastest way to develop a creative and energised atmosphere for the brainstorm group.
- Produce as many ideas as possible; quantity will, in time, produce the quality.
- Requirements engineers should write every idea down, without censoring.
- If the session gets stuck, requirements engineer seeds the session with a word randomly, and asks participants to make word associations that have some bearing on the system to be developed.

In addition to workshops from Case Study 4, brainstorming meetings were effectively used in Case Studies 3 and 4.

8.3.4 Work modelling

Work modelling is actually business process re-engineering of the work. Thus, the greater the scope of the study, the better understanding and better the chance of finding areas that will benefit from improvement. The requirements engineer will see the underlying policy of the users' system by means of work modelling. Work modelling is originally proposed in the contextual design approach by Beyer and Holtzblatt (1998) and was successfully used in Case Study 3 (Arayici *et al.*, 2006). These models are explained below:

The Flow Model: This model defines how work is divided across people and how people coordinate to ensure that the whole job is done. The flow model offers a bird's-eye view, showing the people and their responsibilities and the communication paths between people.

A typical construction project embraces many stakeholders at each phase of the construction project life cycle. The flow model will represent the coordination, communication between the stakeholders and their responsibilities and duties. Because the construction industry is fragmented in nature, the flow models should address this issue in order to recover it in the RE process, which will inspire the flow of value-adding activities through lean thinking.

The Sequence Model: This model supplies the low-level, step-by-step information on how work is actually done and how designers need to make detailed design decisions. A sequence model starts with the overall intent of the sequence and the trigger that initiates it. Then it lists each step in order, at whatever level of detail is collected in the interviews.

The sequence models represent the construction activities in a sequential manner, step by step, and show the progress of a multidisciplinary project by relating the activities to each other in order, which inspires the supply chain in the multidisciplinary project. The output of one activity will be the input of the subsequent activity.

The Artefact Model: Artefacts capture traces of people's work practice (Beyer and Holtzblatt, 1998). Artefacts are the entities that people create or use to help them in work. When people use artefacts, they build their way of working into the artefacts. An artefact reveals the assumptions, concepts, strategy and structure that guide the people who work with it. They are manipulated in the sequence models and passed among people in the flow model.

In its life cycle, the construction industry is a fragmented but also complicated, multifaceted and interrelated chain of activities. In such a complex working environment, the construction stakeholders use some tangible and some intangible artefacts to some extent, due to characteristics and attributes of these artefacts and their interpretation by the construction stakeholders .

Modelling these artefacts will enable the CIE development team to explore the advantages and disadvantages of the artefacts and the interaction between these artefacts. Accordingly, the problems can be identified clearly and the solution can be formulated.

The Cultural Model: Work takes place in a culture that defines the expectations, desires, policies, values and the whole approach that people take to their work. Cultural context is the mindset that people operate within and that plays a part in everything they do. Issues of cultural context are hard to see because they are not concrete and technical. They are generally not represented in an artefact or observable in a single action. Instead they are revealed in the language people use to talk about their own job or their relationships with other groups (Beyer and Holtzblatt, 1998).

The construction industry is as old as the existence of mankind and is heavily influenced by the people's culture. The strong influence of culture adds another layer of difficulty in the implementation of the CIE systems. Capturing and modelling cultural aspects will ease the development of a rational process model with the awareness of culture. Furthermore, involvement of the end users in the development and modelling of the cultural concepts will contribute to overcoming the cultural borders against the implementation.

There are some cultural issues that influence the work: Firstly, *standards and policy*, or their absence define or constrain how work is done (or what can be used or bought); secondly, the *power* and *values* of construction companies are likely to change after adopting the CIE system.

A group's own sense of *identity and* the way in which they work is affected by how they think of themselves. People's *emotions* about how they work include the fear of redundancy in being reprimanded for raising difficult issues or their pride in the work they do. Lastly, the characteristic *style*, *values* and *preferences* of an individual or team are creating a work environment that obstructs other people working in the same environment.

The Physical Model: Any product or system must live within the constraints of the physical environment. The physical environment constrains what people can do, but within those constraints, people do have some control over their environment. Studying the workplace offers important clues to the way people restructure the work to the extent they can. Because they structure their environment to be convenient, the structures they create mirror their thoughts. The structures show which people group together in the same conceptual units and on similar coherent tasks.

The main features of the physical models are indicated in italics. The *places* where construction works take place are rooms, workstations, offices, construction sites, etc. The physical *structures* that limit and define the space are sites, walls, basements, desks, file cabinets, etc. The *usage* and *movement* is about how people move and move artefacts in the space during the work. The *hardware, software, communication lines* and other *tools* in the space support the work.

Network connections are shown in the model to emphasise who is connected to whom and what communication occurs amongst people and what can be automated in the system. The *artefacts* are what those people create and modify and pass around for support. The *layout* of the tools, artefacts, movable furniture and walls are related to each other, to support specific work strategies.

8.3.5 Second version of the requirements specification deliverable

Once the requirements gathering process is completed, the second version of requirements specification deliverable is produced and distributed to all the relevant parties situated in different places. This version updates the first version but also complies with the main goals and aspects explained in the first version. Furthermore, it mainly incorporates raw requirements data collected through interviews, workshops, brainstorming sessions and work models (flow models, sequence models, physical models, cultural models and artefact models).

8.4 Building a Shared Understanding

It is not sufficient to understand the stakeholders' needs. It is also necessary to establish a shared understanding between multidisciplinary users. This shared understanding needs to be developed through conversation and mutual inquiry into the meaningful facts about the end users' work. Subsequently, the different members of the team can learn the perspectives of different stakeholders.

8.4.1 Interpretation sessions

Building a shared understanding commences with an interpretation session. In the interpretation session, the requirements engineer walks through the data collected by means of interviews, workshops, brainstorming sessions and work models. The rest of the team listens to the engineer, asks questions, draws work models, records issues and interprets and designs ideas. In their discussion of what to model and what to record, the team intensively analyses the data and what it means, learns how each team member views the data and develops a shared understanding.

The interpretation session is an efficient way to achieve several desirable benefits such as better data, written record of the end user insight, effective cross-functional cooperation, multiple perspectives on the problem, development of a shared perspective, true involvement in the data and better use of time.

8.4.2 Consolidation

Once a shared understanding starts emerging in the interpretation session, the individual work models that represent the perspective of each end user are brought together, so that the team can establish a common structure and pattern of work across all the end users, without losing any individual variation.

Drawing an affinity diagram is the first step of consolidation. The affinity diagram organises the individual notes captured during the interpretation sessions into a hierarchy revealing common and key issues. That is to say, it reveals all the issues, worries, and the key elements of work practice relevant to the focus of the project team. It also defines the key quality requirements on the system: reliability, performance, hardware support and so on. By reading the affinity, a requirements engineer not only learns the key issues, but can also see the exact data that contributes to identifying each issue in the work (Beyer and Holtzblatt, 1998).

Next, each type of work model is consolidated separately after defining the affinity diagrams. The consolidated models make the underlying patterns of work explicit to the distributed end users. It will now be efficient for the project team to streamline the work, taking advantage of technological possibilities and removing non-value-adding steps, lead times and repetitions. Besides, no list of needs or requirements will give requirements engineers this synthesized holistic view; treating each need or requirement as an independent entity makes it too hard to see how they interrelate.

It is apparent that the same kind of thinking process drives the consolidation of different work models in each category. Information in the affinity diagram about the users is collected and built into groups. Users are assigned roles in the work flow with specific responsibilities. Work steps are collected and grouped into abstract steps in the sequence model. Parts of artefacts are collected, grouped and placed in the physical environment in the artefacts models and physical models respectively. Influencers and influences are collected in one cultural model. Furthermore, understanding intent, strategy, structure, concepts and mindset are keys to effective process and system design.

Discussing each model in turn leads to a synthesis of the issues across models. The team can absorb one coherent aspect of work at a time, making the complexity manageable, and develops a shared understanding of the data and a sense of direction for the system design. Subsequently, work models give the team a handle for both incidental and essential complexities of multidisciplinary activities, encouraging them to respond to the work practice as a whole. They make these aspects of work visible to all the team.

8.4.3 Communicating to the stakeholders

The team members may be situated in large and dispersed organisations. Therefore, some team members cannot be involved in the meetings for interpretation and consolidation. This may be due to discrete reasons, such as the distribution of stakeholders and the part-time involvement of some stakeholders.

Communicating to people who have a stake in the project is part of the requirements engineers' job. Communication provided to inform all external and internal groups must include what the project is doing, details that allow the group to understand the project's design direction and meaningful ways for the groups to comment and contribute knowledgeable ideas. Furthermore, a good communication mechanism also allows for immediate feedback and it will reveal the end users' work practice as a coherent whole.

8.4.4 Third version of the requirements specification deliverable

The third version of the requirements specification deliverable is produced following the completion of the activities for building a shared understanding, including interpretation, consolidation and communication to the stakeholder, whilst the deliverable covers progressive evolution of a shared understanding through interpretation and consolidation. Each consolidated model and affinity diagram should be included and explained in detail with the justifications required. It should also reflect itself as an evolution from the previous versions' deliverables. Furthermore, it should indicate cross-referencing to the previous deliverables for the same issues and aspects to avoid repetition. Lastly, it should be distributed to all the relevant parties, ensuring that every party will be aware of the progress on the user requirements.

8.5 Visioning and Process Modelling (Storyboarding)

Storyboarding, originally used in movie making, is a technique for pictorially building scenario models, and pictorially capturing the new procedure for a task. In Case Study 3 requirements activities, storyboarding was not regarded as a pictorial representation of the new work procedure. In fact, the storyboarding was a text-based scenario. These text-based scenarios were very effective in developing a shared understanding, communication and integration, and conducting an effective testing process. Therefore, storyboarding is regarded as a shared text-based scenario that develops the proposed RE process through visioning. It also forms the basis for the subsequent requirements activities such as system design, use case modelling and prototyping.

As a result, establishing process modelling (text-based storyboarding) will lead to process-led system development, which was stressed in Case Studies 3 and 4. The following paragraphs describe the development procedure of innovative visioning and process modelling (or text-based storyboarding). The goal of visioning and process modelling is to look across the different consolidated models and see a unified picture of work practice, to use the different team perspective to reveal the issues and to use a wide exploration of multiple possibilities to drive the invention of a creative solution for the system design. For example, construction works are complex, multifaceted and complicated, with too many details. Therefore, it is required that the CIE development team should immerse themselves in the work details so that they can see and respond to the work issues together. In each specific consolidated model is put a specific dimension of work into focus for the team by the leadership of the requirements engineers and managers. Each model reveals problems and issues related to dimensions of construction activities.

It is important that the project team thinks widely and considers several alternatives, including addressing radical solutions to point out different aspects of the work situation. These different solutions are consolidated into one response, which incorporates the best ideas into a single unified corporate response. To be successful, this corporate response has to be tied to reality. It has to fit with the end user's work in detail.

A good process designing will define explicit steps for process modelling. The contextual design technique used particularly in Case Study 3 provides a set of steps linked together into a process that moves the team to a concrete representation of their shared vision. These steps are walking the data, visioning, evaluation and integration and concrete action. These steps are briefly explained below.

8.5.1 Walking through the data

Walking the affinity diagram and consolidated work models focuses the team on specific aspects of work. Each model will generate a set of goals; values to encourage; negative feelings to eliminate; roles and activities to support, combine, or eliminate and so on. After walking the affinity diagram and work models, the team thinking is crystallised by making a list of the most important issues from the models. When the lists are made, it is time for visioning.

8.5.2 Vision development for business process redesign

Visioning is done through brainstorming sessions. Ideas for visioning are not evaluated as they are generated in the brainstorming session because ideas are driven by the end users' work practice. Brainstorming sessions for visioning were successfully carried out in Case Study 4.

A visioning session gives the team a chance to choose a starting point and spin it out into a story about the new work practice transformed by technology. The story describes the new work practice that the team envisions. A vision session starts with one of the ideas from the list of starting points and incorporates each idea into a coherent story about the redesigned work. It shows how people are in the roles they play, the IT systems they use, how they communicate with each other and the systems and how the systems are structured.

8.5.3 Evaluation and integration for shared vision and process model

Visioning through brainstorming sessions may result in multiple visions, each of which is built by the whole team and incorporates everyone's different perspective, each suggesting a different design direction or addressing a different part of the multidisciplinary work.

The key to such a design is to treat each vision, not to accept or reject it as a whole; every vision has impractical or undesirable elements and some other elements that cannot be lost, creating a better solution, by identifying elements that work, recombining and synthesising them by preserving the best parts, and extending them to address more of the multidisciplinary work eliminating any defects. The whole process is designed to bring a disparate, cross-functional team of people to a consensus and a solid shared understanding.

8.5.4 Technical action for technological possibilities

The process model defines what is expected of any software and hardware components of the system. Engineers can have an advanced look at what demands are made on technology. They can start investigating technological possibilities immediately, including possible platforms as to whether specific technology is sufficiently reliable and whether it can meet the requirements of the vision and user interface possibilities.

In the end, the process model contains a number of scenes that tell a detailed story of a specific piece of construction duty. The model is used for planning how the story will develop as each scene reaches its outcome. Furthermore, it will form the basis for the system design and modelling in the subsequent stages.

8.5.5 Fourth version of the requirements specification deliverable

At the end of the process modelling stage, the fourth version of the requirements specification deliverable is produced, which represents a process model (or text-based storyboard) as an output of the visioning and process modelling phase. It should reflect the evolutionary progress of the RE process and it should explain how the process model is derived from the consolidation models and developed through visioning. Furthermore, it should also comply with the previous deliverables and address the updates and changes in requirements with cross-referencing to the previous deliverables. This is required for each stakeholder to follow the requirements changes and traceability and to strengthen the shared understanding amongst the team.

8.6 System Design

System design is the stage where the user requirements are transformed into a system design, which was the issue addressed in Chapter 5 as reconciling software requirements and architecture. The challenge is to keep the system design coherent without missing any single requirement, so that it supports the users and fits with their expectations, while extending and transforming their work practice prescribed by the vision and process model into the system design.

The User Environment Design (UED) technique of contextual design approach can be adopted for transition from storyboarding to system design. The UED plays the same role that the floor plan plays in a house design. A floor plan represents the key distinctions to support living. Similarly, the UED represents the key distinctions to support work practice with software systems. These key distinctions are called *focus areas* (Beyer and Holtzblatt, 1998) that show the coherent places in the system doing an individual activity for the work. They are the rooms of the system, like the rooms in the real world; they should support activities that occur in the rooms. Each focus area has a purpose that is a concise statement of the work that the focus area supports. Furthermore, the focus areas should provide the functions that are required for the work in the focus area. They should contain, organise and present the objects that users need to work on. They should also include the links, constraints, issues and so on. Shortly, because focus areas are the most visible concept captured by a user environment model, they help the requirements engineer organise the system.

UED is placed in the requirement engineering process between the process modelling and use case development with Unified Modelling Language (UML). UED bridges the natural rotation between sequential and structural thinking. Storyboarding and use cases are sequential. They indicate a single series of events in order. The UED and object models of UML are structural. They show all parts of the system and how they interrelate with the system work model and internal structure.

Process modelling drives the design of the user environment because storyboarding, which is sequential, includes threads for the system design but UED is structural and reveals how all the threads fit together coherently. The UED enables the building of the single coherent structure that supports multiple tasks performed by different roles in a multidisciplinary work environment. Since the system is a mix of hardware and software, the focus areas in UED represent physical hardware places as well as software. Hence, UED diagram can be extended to represent the total system delivered to the user: hardware, software and documentation (Beyer and Holtzblatt, 1998).

8.6.1 User environment design walk-throughs and inspections

The walk-through and inspections are to ensure that the team, including the end users, are clear on what they intend to create by the design and how they think it will work. It enables the requirements engineers to identify the test cases and conditions that will be a test focus of the users. The questions to ask for checking a UED recommended by contextual design (Beyer and Holtzblatt 1998) are as follows:

- **Are focus areas coherent?**
  The project team should check that each focus area supports one activity within the overall task and that the title and purpose statement represent it. The project team should suspiciously approach the focus areas that have no purpose; this is because the team is not clear on what the purpose is.
- **Do focus areas support real work?**
  Focus areas group the related functions, but may not support something that the end user works on. Focus areas that do not support a coherent work task, but instead only reveal the data associated with an object in the system, are inspected.
- **Are functions correct?**
  Functions that are not in direct support of the focus area are inspected. These functions may imply a distinct activity that should be separated into another focus area.
- **Are focus areas distinct?**
  It is required to collect the focus areas that support the same part of the work, the same activity, task or role, and compare them to see if they are clearly distinct; if not, they may be required to recombine in order to give a cleaner set of distinctions for doing the work.
- **Do links make sense?**
  The links should support the work task. Certain patterns of links and focus areas may appear or indicate problems in the design.
- **Is the work supported?**
  Consolidated models and the process model (storyboard) are used to refresh memory, and the UED is scanned through from the point of view of the different roles and tasks. The project team makes sure that the design works for different kinds of users. It should account for the issues the end users care about.

8.6.2 Fifth version of the requirement specification deliverable

This version can be called *software specification* as well as requirements specification because it will explain the outputs of the UED phase, which defines the specifications of the system design too. UED will explain how the new system will behave and organises its function in a way that makes sense for the user. The deliverable should contain the following sections:

Section 1: Overview: The first part gives an overview of the whole system, its goals and its business structure with referencing to the previous deliverables.

Section 2: Supporting data: This section summarises the end users' data on which the system design is based. It should have a coherent association with the previous deliverables and reflect the evolutionary progress of the RE process. It is important to emphasise how a design is built and how it responds to end users' requirements.

Section 3: Functional requirements: This is the basic definition of what the system does. It is organised by focus areas in the UED. Each section of functional requirements introduces the focus areas and describes the work done in the focus areas. Each function is named and defined with a full description of the function's behaviour.

Section 5: Non-functional requirements: Additional requirements on the system such as performance, reliability, maintainability, usability, platform supported, etc. are listed. These are collected from the affinity, and extended while building the UED but they are not associated with any focus area.

Section 6: Objects: The objects manipulated in the different focus areas are listed with the focus areas but described once. The meaning and usage of objects across all focus areas are described. These objects have actual associations with the artefacts explained in the artefacts modelling.

Section 7: External interfaces: In this part of the deliverable, external interfaces to the system are collected and described.

8.7 Use Case and Object Modelling with UML

The UED does not focus on identifying common objects as its primary feature. Therefore, it is clearly not object oriented. Furthermore, it is not task oriented because it prescribes no order, unlike the process modelling and storyboarding. It shows parts of the system and their relationships in a structural manner.

In fact, object-oriented design, from use cases and object modelling approaches, does not stand on its own. The purpose of a use case is to tell a coherent story of how the users will work and how the system will meet their needs. Object modellers can extract the objects and their behaviours. However, neither use case nor object modelling provides a good representation of the system work model. The use case is task oriented, telling one story of use for one task, but it does not provide a coherent view of the system.

On the other hand, object models give a coherent view of the system, but not what the system user experiences. Instead, it is a view of the system that the developers will experience and implement. It is not user oriented because it is not focused on keeping the implementation coherent, extensible and maintainable (Beyer and Holtzblatt, 1998) for stakeholders of a multidisciplinary work environment such as construction.

The main question is how to decide on what objects and behaviours are to support the users. That is the question answered by the UED. Instead of basing the object definitions on use cases, the UED provides a coherent model of the system that can be designed and corrected before object definition starts.

8.7.1 Implementation of the system design

The next task, after developing the UED, is to start the implementation. This is where use cases and object modelling take place. The design work through process modelling and UED will give the team the foundation to structure the implementation through use case and object modelling, because the UED development is still insufficient to model the solution for the problem. In other words, storyboarding and UED are efficient techniques to model the problem and to express the requirements and constraints on the system to be developed. After the problem is modelled, it is time to formulate a model of the solution.

If the model of the problem is too far from the model of the solution, a great deal of effort may be expended in attempting to translate the expression of the problem from a form understandable to the end users into a form understandable to designers and programmers. This implies that there are many opportunities for misinterpretation and often makes it difficult to validate the solution with respect to the stated problem.

An effective method recommended by the rational unified process is the technique of use case modelling. Use cases provide a means of expressing the problem in a way that is understandable to a wide range of stakeholders: end users, developers and any other interested stakeholders. The transition from process modelling (problem modelling) to UED (solution modelling) and from UED to use case modelling (problem modelling) and from use case modelling to object modelling (solution modelling) will ensure that the problem modelling can fully match the solution modelling in the whole part of the CIE system. Each use case is a complete flow of events in the system from the user's perspective. Moving from use case to object model is another example of switching between sequential and structural thinking. The use cases are a story of use, the object model is structural and object interaction diagrams are another story. However, as practised by the engineers and researchers, such as in Case Studies 3 and 4, use cases may be of a very high level, showing a whole task in the work, or low level, showing the accomplishment of a smaller function. Added to this, it is always an issue to decide what should be specified by a use case.

In the RE approach being proposed, that is aligning with the contextual design at this stage, integration with use cases at two levels for a coherent synthesis is ensured. Firstly, process modelling acts like high-level use cases. They show how real users interact with the system for their duties. At this level, the process model is well grounded in a consolidated model and vision, so there are clear criteria for what should be included. The stories of tasks in the process model make it easy to scan and see the emerging design. Furthermore, while use cases include preconditions, post conditions and exception, it is not necessary to specify the use cases at the process modelling level.

The UED provides a high-level structural thinking step that responds to the process model, whereas use case models, which are developed based on the preceding UED development, will also provide low-level sequential thinking that responds to the UED.

Object modelling captures the implications by switching back to structural thinking from low-level use cases. At this level, each use case tells the story of how one function or a closely related group of functions operates. The use case is based on the process model and the UED: the storyboard defines what the user will do and the UED defines the functions. The use case works out precisely what happens when the user operates these functions, how the system responds and how the system internals make the designed response possible. Object modelling with UML bridges the gap between the system design and the system implementation.

Building the UED as an intermediate step between process modelling and use cases ensures that the structure built into the use cases holds together for the user. Until the user environment structure is stable, building the use case modelling should not commence because changes at the user environment level will affect what happens in the use cases. The more revealing, identifying, addressing and testing of the structural issues with the end users before starting the use case and object modelling, the faster are the modelling and coding carried out.

8.7.2 Sixth version of the requirements specification deliverable

This deliverable will have a structure similar to the fifth version, but it will include more precise and detailed specification than the previous deliverable and denotes a software specification deliverable, rather than requirements specification. However, it is the extension of the fifth version and further explains the results of smooth transition from the user requirements to the system design and modelling.

Use case models and object models developed at this stage are explained in detail with referencing to the previous deliverables for the corresponding aspects identified in the process modelling and the UED.

8.8 Incremental Prototyping with the End User Tests as an Agile Process

Testing can be described as the mechanism for assessing the quality of the end product and for strengthening and stabilising the architecture early in the development life cycle and for determining whether it matches its requirements specification and executes in its intended environment. Iterative prototyping with end user testing can be regarded as an agile process for analysis, verification and validation of the requirements, as well as a way of achieving the quality by eliminating the errors in the system. It requires intensive user involvement, non-functional requirements, process awareness, iterative and incremental improvement, intensive teamwork and collaboration and a high level of systemic thinking (see Chapter 4). This iterative prototyping through end user tests was successfully carried out in Case Study 3.

The purpose of testing is to assess product quality, which not only encompasses the assessment of the final system before deployment but also embraces the assessment of the requirements and system architecture and starts early in the project and continues through the assessment of the final release delivered to the end user. Although the RE activities with use case modelling and contextual design approaches were conducted in an ad hoc manner in Case Study 3, the agile process of incremental prototyping enabled the project team to establish a coherent and consistent association with the contextual design and use case-driven analysis techniques.

The use case modelling and contextual design approaches enabled the project team to update and validate the user requirements. However, if the early requirements activities before coding were conducted in a systemic manner, the effectiveness of the incremental prototyping would considerably increase. This is because the required material and documentation for the incremental prototyping are provided from the requirements deliverables of the preceding requirements phases such as process modelling, end user environment design and use cases.

Reverse engineering can also be conducted to match the code with the object models. However, this can be more beneficial to the developers than to the end users because reverse engineering involves white-box testing, which is concerned with the internal structure of the system (Whittaker, 2000). On the other hand, the tests of the incremental prototyping focus on testing the system against the written specifications from process modelling, UED and use cases. In addition, the tests observe the system components as black boxes and are totally unconcerned with the internal structure of the programmes.

There should be a symbiotic relationship between the end users and the developers. The entire set of tests in the life cycle should be conducted by the end users, not the developers, through the initiation of the requirements engineers designated as testers. The developers release the early prototypes and the end users, who can be distributed across different places, test the prototypes continuously in a collaborative manner with respect to the defined testing objectives. They continuously provide feedback to the developers throughout the test phases.

In Case Studies 3 and 4, the main focus was on the functional requirements in conducting the requirements activities with the use case-driven analysis and the contextual design approach. However, the incremental prototyping entailed taking the non-functional requirements into consideration such as usability, performance, reliability and so on.

Incremental prototyping in Case Studies 3 and 4 involved the demonstration of the system to the construction practitioners while the system is evolving in the development life cycle, in order to introduce the CIEs to the construction industry, which helps them have a positive vision for the future to exploit such IT in order to create the lean construction environment.

Furthermore, continual involvement of the users in the prototyping is very important to gain their trust, because they can see the progress, which they can communicate to the developers, and the way their responses shape the system. As a result, the interest and involvement generated by the demonstration sessions of the subsequent releases of the early prototypes in the development leads to easier acceptance and adoption of the system to the industry.

Since the UML testing workflow was used in the incremental prototyping and it positively affected the user requirements activities, the same approach is adapted to the RE process being proposed. A UML test workflow that was successfully employed in Case Study 3 particularly encompassed the stages: plan test, design test, implement test, execute test in the integration tests stage and execute tests in the system test stage. These stages are already explained in Chapter 6. However, a coherent association between the agile process and the other techniques such as storyboarding, UED and use case modelling is required. In the following sections, the attempt for this is explained.

8.8.1 Test plan and design

The deliverables produced from the process modelling, UED and use case modelling stages provide the information necessary to develop test plans. The process modelling shows how the system should work, the UED provides the formal definition of the functions and use case models enable the user to realise these functions in the process model. It is straightforward to build a test plan that will check the requirements specification against the actual system. In addition, the test plan produced before the test is the seventh version deliverable of the RE process. Before testing starts, it should be distributed to all the interested parties in the project team and there should be a consensus established between the stakeholders.

The purpose of the test plan is to identify and describe the testing that will be executed. Initially a single test plan is prepared, describing all types of tests to be executed for successive prototypes at the subsequent stages in the project and then the test plan is modified at each testing phase. Based on the test plan, the test efforts were measured and managed. On the other hand, the purpose of the test design is to identify, describe and generate the test model. The test design transforms the requirements specifications into alpha, beta and final test cases. The outputs from the use case modelling, UED and process modelling are utilised for the test design of alpha, beta and final tests respectively, because they drive the design of the system elements that implement the tests.

8.8.2 Alpha phase testing (unit test)

Alpha testing phase is the initial phase of the testing process and it denotes the unit testing. In other words, system components or the applications that make up the system are tested as stand-alone by all the end users situated at different locations. The system components are tested against the corresponding use cases that encapsulate different requirements. As stated by Stevens and Pooley (2000), use case modelling helps with three of the most difficult aspects of development, which are as follows:

- **Capturing requirements:** Use cases are linked to the UED and process modelling in the RE process for requirements capturing and modelling.
- **Planning iterations of development:** The UML object models are derived from the use cases to draw up the details of system architecture in line with the user requirements. Lastly, it bridges the gap between the system design and the object modelling and coding.
- **Validating systems:** Each use case describes a requirement on the system so that a correct system design should allow each use case to be carried out. An obvious and very useful technique for validating a system design is to take each use case in turn for testing.

In Case Study 3, the main objectives of the alpha phase testing were the functionality and usability testing. Each product's conformity was observed and examined against the user requirements through functionality testing. Based on the experience and lessons learnt from Case Studies 3 and 4, the alpha phase testing is embedded into the RE process being proposed.

The advantage of alpha testing is that it enables the distributed end user groups to test the individual components and explore any possible defects belonging to the applications and to record them and inform the developers accordingly. Then the developers improve the system components accordingly for a better way to manage integration to form the whole system at the subsequent phases.

At the end of the alpha phase testing, the requirements engineer, who initiates the testing, collects the test results, the seventh version deliverable is produced and the overall results are distributed to the relevant parties. Furthermore, the deliverable should also include the updated test plan and strategies for the subsequent testing phase, which is the beta testing phase, so that all interested parties will be aware of what will take place next.

8.8.3 Beta phase testing

Beta testing is related to integration testing. The test cases are prepared from the UED outputs. In other words, beta testing is not only concerned with functional and non-functional requirements; the goal is to ensure that there is a basic skeleton system (system architecture) that will serve as a test platform.

Activities undertaken by the different construction practitioners in the focus areas and the links between the focus areas for exchange of information amongst the practitioners are tested. Furthermore, using the UED outputs will enable the practitioners to test the communication between the distributed parties situated at different locations. Beta testing is the initial phase of the integration to form the system of CIE.

Since use cases are derived from the outputs of the UED, the project team can also fully retest the whole functionalities of each system component, which are already tested at the alpha phase. This is useful for observing the expected improvements from the alpha and beta tests in functionality and any defects encountered at the alpha-testing phase undergo the regression testing in which the defects have been supposedly removed at the beta phase (Whittaker, 2000). Furthermore, new defects discovered at the beta phase were also logged into sheets.

At the end of the beta phase testing, the initiative tester collects the test results and the eighth version deliverable is produced; the results of the beta phase testing are distributed to the end users and the developers to improve the system accordingly. The deliverable should also include the updated test plan and strategies for the next testing phase, which is the final tests, so that all interested parties will be aware of what will take place next.

8.8.4 Final phase testing

At this stage the whole system is tested in many respects. The planning for system testing starts with the test cases, which are generated from the text-based storyboard (process model) and are sufficient to demonstrate the whole system from a black-box perspective. Final phase testing is a multifaceted test to assess the operation of the whole system in functionality, reliability, performance and usability. System testing tests the collection of the applications that constitute a deliverable system and the entire domain is considered to satisfy the criteria for a system test (Whittaker, 2000).

The process model, which is developed through visioning, corresponds to the construction supply chain. Therefore, system testing at this stage relies on the process model, which entails the interaction of the components, exchange of information between the distributed stakeholders and a synchronous or asynchronous type of collaboration. Issues tested at the beta phase are also retested at the final phase. Because the UED is derived from process modelling, using the process model at the final phase will enable the project team to review all the issues from the beta phase.

Final phase continues iteratively until the end users are satisfied with all the test criteria because the final phase testing also covers the issues of acceptance of the end users, which is defined as the reaction of the stakeholders to the system and their overall evaluation on how much the system impacts their business life. This is critically important for the implementation of the system in a multidisciplinary industry such as construction. To ensure the acceptance from the end users, industrialist should be involved in the testing life cycle, in particular in the beta and final phases. Consequently, the system will have what the industry wants in real construction life.

At the end of the final phase testing, the requirements engineer, who initiates the testing, collects the test results, and the ninth version deliverable is produced; the results of the final phase of testing are then distributed to the relevant parties. In the final version, user reactions, validation of the requirements and the system and exploitation of the system in the industry are addressed. Consequently, it is expected to be a successful system development through a consistent and effective RE process.

8.9 Summary of Mastering the Requirements Engineering Process

The RE methodology being proposed is targeted at CIEs for developing collaborative, distributed integrated information systems for multidisciplinary work environments. The key requirements for the method are that they should be (1) ready to use, (2) simple, (3) domain specific, (4) adaptable, (5) systematic and (6) integrated with the legacy systems.

The method itself consists of some key constructs, such as an interrelated requirements document deliverables for requirements specification, techniques for the RE process, process and product awareness for requirements management and tool support such as UML. There are two reasons to address all the constructs in the same method. Firstly, it will be possible to use the same method to support the whole software life cycle. Secondly, it makes it possible to start using the method in the CIE developments even without any adaptation.

The framework is depicted in Figures 8.1 and 8.2.

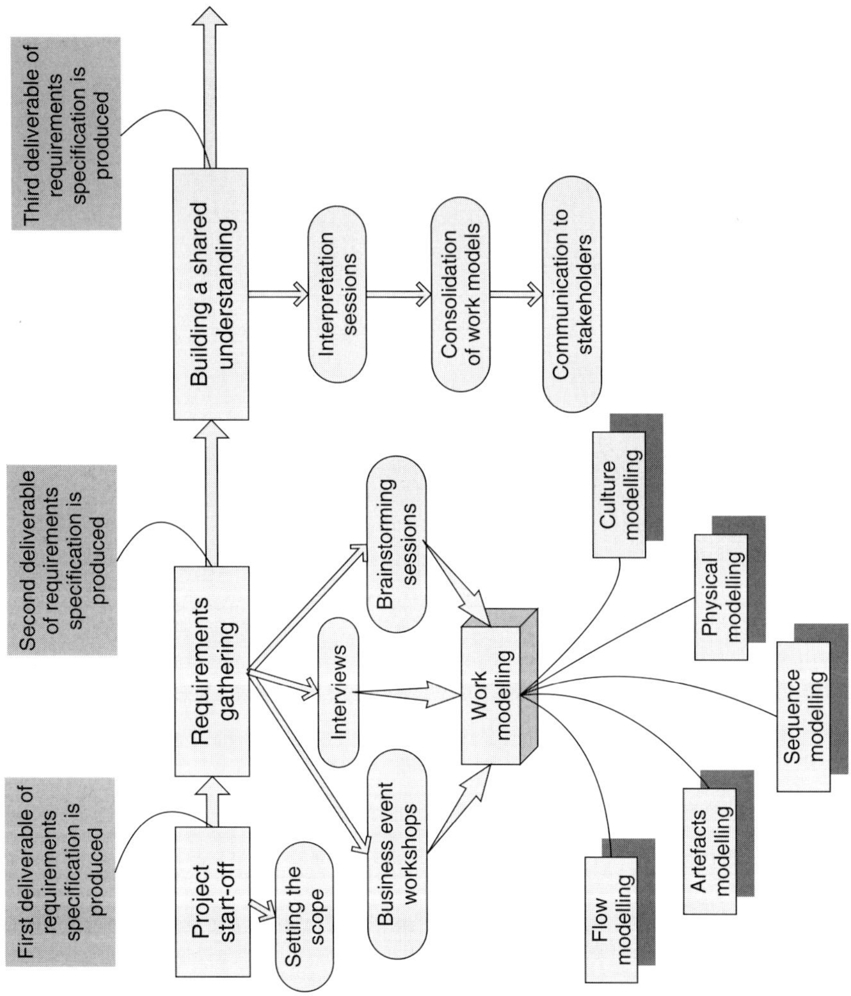

Figure 8.1 The overall framework of the requirements engineering process (first part).

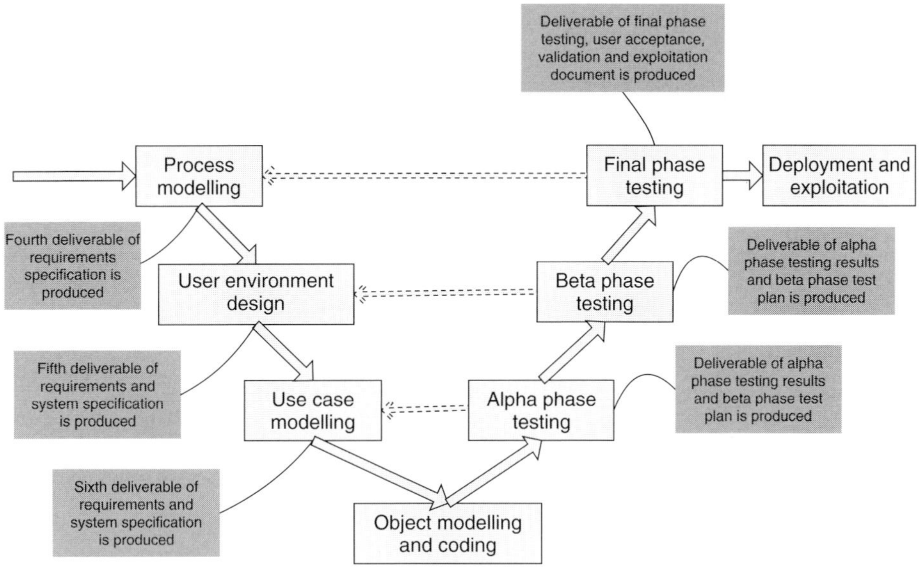

Figure 8.2 The overall framework of the requirements engineering process (second part).

References

Arayici, Y., Ahmed, V., Aouad, G., (2006), "A Requirements Engineering Framework for Integrated Systems Development for the Construction Industry", Electronic Journal of Information Technology in Construction (ITcon), 11:35–55, www.itcon.org/2006/3.

Beyer, H., Holtzblatt, K., (1998), "Contextual Design. Defining Customer-Centered Systems", San Francisco, California, Morgan Kaufmann Publishers.

Christiansson, P., (2001), "Capture of User Requirements and Structuring of Collaborative VR Environments". In O. Tullberg, N. Dawood, M. Connell, editors, AVR II & CONVR 2001. Conference on Applied Virtual Reality in Engineering and Construction Applications of Virtual Reality, Gothenburg, pp. 1–17, 201pp.

Stevens, P., Pooley, R., (2000), "Using UML Software Engineering With Objects and Components", Updated Edition, Edinburgh Pearson Education Limited.

Whittaker, J.A., (2000), "What is Software Testing? And Why Is It So Hard?", IEEE Software, 17(1):70–79.

9

Evaluation of the Proposed Requirements Engineering Framework

9.1 Introduction

The evaluation and assessment of the proposed requirements engineering (RE) framework for computer integrated environments (CIEs) is possible in two ways, empirical and/or theoretical methods; however, empirical assessment of the framework is not within the remit of this book, as it will require another development and implementation project, where the proposed RE framework is adopted as a unit of analysis. The implementation and testing of the framework being proposed in this book, either on another CIE system development or on building information modelling (BIM) implementation in design and construction companies will require another 2 or 3 years' work at least, as well as enterprise activities. However, it is strongly recommended to implement and adopt the framework in the future CIE developments and BIM implementations to include both research and commercial projects.

Theoretical evaluation and assessment of the framework is possible and is done in two steps. In the first step, the framework is to be assessed against the key issues and criteria explained in Chapter 5, based on which the RE practices in Case Studies 3 and 4 are analysed and improved in Chapters 7 and 8 respectively. In the second step, it will be assessed against the external or independent evaluation and assessment models, which are not used in the development of the proposed RE framework in Figures 8.1 and 8.2.

9.2 Internal (Dependent) Evaluation

In order to conduct the internal evaluation, the proposed RE framework is assessed against either the list of criteria or the key issues explained in Chapter 5. The list of criteria is actually designed to

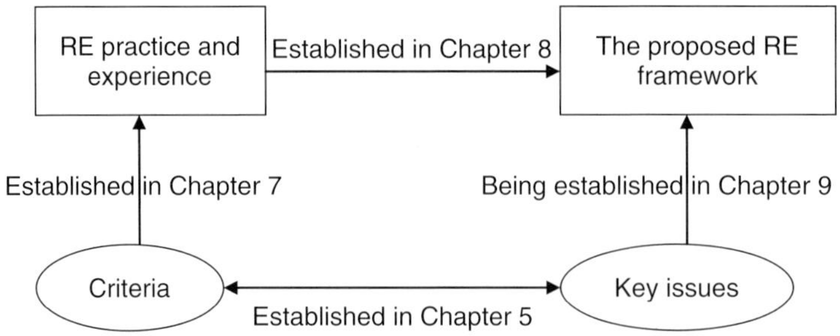

Figure 9.1 The coherent associations between the four aspects in the development of the proposed RE framework.

measure the success of the RE processes, after the implementation of the RE approach in a system development project. Therefore, these criteria have been used for the analysis and evaluation of the RE practices mainly in Case Studies 3 and 4. However, using the same criteria for a proposed RE process will not be appropriate because it has not yet been implemented. On the other hand, the key issues were specified for evaluation and assessment of the RE processes before the implementation of the RE processes.

As a result, using the key issues for the internal evaluation of the proposed RE framework is more appropriate when compared to the criteria. In addition, making use of the key issues as a tool for evaluation and assessment will make the associations more solid among the four aspects, which are the key issues, the criteria, the RE activities in Case Studies 3 and 4 and the proposed RE framework. The association between them is depicted in Figure 9.1.

The figure also denotes the evolutionary development of the proposed RE framework. Basically, as explained in Chapter 5, the association between the key issues and the criteria was established through benchmarking. After the critical analysis and elaboration of the RE practices in Case Studies 3 and 4 in Chapter 6, another association between the criteria and the RE practices was established in Chapter 7 to evaluate, analyse and measure the success of those RE practices through a survey. Through the interpretation of the findings from Chapters 5, 6 and 7, the RE practices from Case Studies 3 and 4 were further improved and mastered in Chapter 8 to propose the RE framework as illustrated in Figures 8.1 and 8.2. Lastly, the attempt in Chapter 9 is to establish a final association between the key issues from Chapter 5 and the proposed RE framework to complete the loop for internally consistent evaluation and assessment of the proposed RE framework.

This internal evaluation seen in Table 9.1 is conducted by assessing the proposed RE framework against the key issues. The similar key issues extracted from different themes in Chapter 5 are put together in the same rows in the table below for concise evaluation.

All in all, the proposed RE process covers all the key issues and it also supports and facilitates the key issues that are related to requirements management such as conducting requirements management, a cultural impact on the RE process and increasing the maturity of the stakeholders by requirements management activities. As a result, the proposed RE framework is validated according to the internal evaluation.

9.3 External (Independent) Evaluation

External evaluation is conducted in two steps. In the first step, the proposed RE process is assessed against the top ten guidelines of the Requirements Engineering Adaptation and Improvement for Safety and dependability (REAIMS) assessment model and in the second step, requirements-related project risk factors (see Chapter 2) are used for the evaluation.

The concept of maturity is a good means of describing and improving the quality of the practices in system development (Nikula, 2003). The Capability Maturity Model (CMM) especially provides a good reference point for maturity evaluation. The REAIMS maturity model has three levels and its assessment includes evaluation of 66 good RE practices. However, the top ten criteria of the REAIMS assessment model should be adopted for RE practices in order to make sure that these RE practices achieve level 1 maturity.

The following is the evaluation validation of the proposed RE process with the REAIMS top ten list. Since the original REAIMS assessment addresses both the practices and their usage patterns, calculating maturities is possible before the proposed RE process is applied in practice, preferably in multiple projects. However, theoretical evaluation of the proposed RE process with the top ten guidelines, which are indicated in bold in Table 9.2, is possible.

The second step of the external evaluation is conducted with the requirements-related project risk factors. Eight project risk factors are listed in Chapter 2. These factors are used in Table 9.3 together with the techniques that address the factors in the proposed RE process. The proposed RE process first tries to eliminate the likelihood of these risks, and secondly it should make the presence of these risks apparent so that appropriate actions can be taken.

Table 9.1 Internal (dependent) evaluation.

Key issues	The coverage of the key issues
K1,K2,K3,K4,K14,K23	*Traceability in RE process*

Traceability is adopted in the proposed RE process, particularly following the requirements elicitation. Some tasks in the proposed RE framework such as building a shared understanding, process modelling through storyboarding, use case modelling and product modelling through the User Environment Design (UED), and object modelling are carried out by means of traceability. The deliverables are produced after subsequent stages to enable the traceability through the stages of the proposed RE framework.

The coherent integration and associations, without tight coupling amongst the stages of the proposed RE process, enable the continuity and continuous traceability. Furthermore, the interrelation and cross-referencing between the subsequent deliverables of the requirements specification helps the project team manage the requirements changes and versioning for traceability. Lastly, communication with the stakeholders also enables everybody to trace the RE process from the outset to the end.

K5,K15,K37,K40	*Handling the complexities in the RE process*

While requirements are initially elicited through interviews, workshops and brainstorming sessions, the five different types of work models are also used to start formalising these informal requirements of stakeholders working in a multidisciplinary environment. Throughout the subsequent stages of the proposed RE process, the formalisation of the requirements through transition between the sequential thinking and structural thinking continue to evolve and handle the possible incidental and essential complexities.

The main characteristics of the subsequent stages of the proposed RE process are described in the framework. A good awareness of the features at each stage and treatment of the evolving requirements accordingly will enable the project team to handle the complexities in the RE process, and this awareness will also enable all the stakeholders to follow and understand the progress of the evolving requirements. Subsequently, their contribution to the development will be effective and satisfying. In addition, developing work models in different dimensions – flow, sequence, artefact, culture and physical environment – allows the team to organise and scale the possible incidental and essential complexities from the beginning in order to handle them well.

K6	*Conducting requirements management*

Because the proposed RE process requires teamwork and collaboration for an efficient and productive RE process, requirements management is conducted by the requirements engineer in collaboration with the rest of the team, including the end users involved in the project from the beginning.

Managing the requirements and processing these requirements throughout the RE process will require communication to the stakeholders, structuring the requirements information with process and product awareness. The coherent and well-balanced associations and transitions for process and product modelling

Table 9.1 (*Continued*)

<table>
<tr><th>Key issues</th><th>The coverage of the key issues</th></tr>
<tr><td colspan="2">in the proposed RE process make it possible to align the requirements management with product and process awareness, as long as the team follows the recommended process to produce the specification deliverables as prescribed and spreads them to all the interested parties.

In short, the proposed RE process will facilitate the requirements management with product and process awareness. Requirements documentation is important to conduct the requirements management activities such as unique identification of requirements, versioning and change management. However, any tool for requirements management is not prescribed in the proposed framework.</td></tr>
<tr><td>K7, K8,K22</td><td>Measuring the RE process for quantitative analysis</td></tr>
<tr><td colspan="2">There is no tool or technique introduced in the proposed RE process for measuring and scaling the development to characterise the elements of the process and product models. However, end user involvement and continuous communication and collaboration with the end users will inevitably require the approval of the end users for an acceptable level of scale for each requirement. In other words, the end users, to accept the requirements at a satisfying level of accuracy or quality, will spontaneously and inevitably identify a fit criterion for each requirement.

With regard to non-functional requirements, though these requirements evolve from work modelling at the requirements elicitation phase, they are not explicitly addressed until the UED, when the non-functional requirements are also modelled. However, they are still not measured at this stage until the incremental prototyping, when the end users test the prototypes with respect to the non-functional testing criteria, when they describe the accuracy and the quality they experienced in the testing and the accuracy and quality they wish to have according to the nature of the requirements.</td></tr>
<tr><td>K9,K12,K17,K26</td><td>Completeness of the requirements specification</td></tr>
<tr><td colspan="2">The deliverables produced after the completion of each stage in the framework form the requirements specification, which actually comes to fruition at the end of the UED, and a flexible template for the requirements specification is identified. However, it does not mean that the previous deliverables are outdated when the RE process reaches the completion of the UED. This is because the emphasis of each deliverable is different and they are interrelated to keep each of them simple and concise as well as to maintain the traceability and versioning of the requirements.

The RE process implies the continuity and coherent links and association between the subsequent activities. Likewise, the subsequent deliverables of the requirements specification have similar associations between them and furthermore, because they are the outputs of the subsequent stages of the RE process, and based on each other, they reflect the continuity in the proposed RE process.

While the repetition is avoided in the consecutive deliverables by cross-referencing, in order to keep the deliverables simple, concise and short,</td></tr>
</table>

(*continued*)

Table 9.1 (*Continued*)

Key issues	The coverage of the key issues
requirements change, and versioning and updates are addressed in the deliverables, together with cross-referencing to the previous versions of the corresponding requirements in the previous deliverables. All deliverables after the subsequent stages in the proposed RE framework provide the completeness and unambiguousness in the requirements specification.	
K10,K13,K29,K35	*Consistency and quality of the requirements*
Keeping the requirements pertinence and discarding irrelevant requirements are also taken into account in the proposed RE framework. The progress in each phase is based on the outputs of the previous phase. Coherency, consistency, connectivity, relevance, traceability and quality are looked for in every phase by means of the specification deliverables, reviews, inspections, walk-throughs and so on. Furthermore, these issues and any misunderstood requirements are checked and corrected during the incremental prototyping, where it is very unlikely that any irrelevant and inconsistent requirements will be missed.	
K11,K18,K32	*Building a shared understanding of the user requirements*
Building a shared understanding is one of the biggest focuses of the proposed RE process. After collecting the requirements, the main purpose of the requirements analysis and modelling phase is to progress on the RE process, while concurrently establishing and improving the shared understanding. The shared understanding emerges through interpretation sessions, consolidations, affinity diagrams and communications to the stakeholders, and evolves through visioning, process modelling and the subsequent stages, with a consensus among the project team, including the internal and external end users.	
K16,K36	*The fit between the structure of the world and the requirements model*
This is also adopted in the proposed RE process. The work models capture the real and current work practices in respect to work flows, sequence, artefacts, culture and physical work environment. In order to redesign the business process through process modelling and system design, which will also subsequently fit the real world, collaboration with the end users and current work practices are improving in line with the project goals and objectives. Furthermore, continuous involvement of the stakeholders will make the requirements fit or compromise the real world according to the level of shared understanding.	
K19,K25,K28,K41	*The Quality of requirements specification*
The quality of the requirements specification is also taken into consideration. In order to make the specification deliverables simple, readable and concise, the structure and context, or focus of each deliverable can be different, but they are coherently interrelated. For example, the first version of the deliverable produced after the project start-off stage embraces high-level project issues such as goals, cost–benefit analysis, scope, stakeholders and so on. However, this deliverable	

Table 9.1 (*Continued*)

Key issues	The coverage of the key issues
does not tell us about the functional or non-functional requirements, which are addressed in detail in the fifth version, in which high-level project issues such as goals, cost and scope are not elaborated, but the fifth version refers to the corresponding deliverable for these issues. On the other hand, having the same standard template for all deliverables of the specification and repeating the requirements-related issues in each deliverable would increase the size of the specification. It would be too complicated for the stakeholders to try and recognise every issue in the deliverables at a time when many stakeholders, who may be involved in such CIE developments, are not very familiar with the RE issues in the CIE system development. Unlike having a standard template for all specification deliverables, it is more beneficial to prescribe the nature and features of each of the specification deliverables. This will then address the discrete outputs of the corresponding stages of the RE process, and will enable the end users, who are actually the construction stakeholders, to envisage and follow the development step by step, and provide good contribution at relevant stages.	
K20,K24,K27,K34	*Thoroughness given to non-functional requirements*
Unlike the previous phases, non-functional requirements are explicitly expressed at the UED. Since functional requirements and the system architecture are considered together at the UED stage, the quality of these functional requirements (non-functional requirements) is inevitably considered explicitly at this stage. In other words, the conceptual differentiation of functional and non-functional requirements occurs at the UED and continues throughout the rest of the RE process. Constraints and cost-related issues are already taken into consideration from the beginning – the project start-off stage – and any updates and changes are conveyed in the subsequent deliverables of the succeeding phases of the proposed RE process. In addition, the level of balance between the functional requirements and non-functional requirements is established at the UED stage, and the non-functional requirements are also modelled at this stage.	
K21	*Thoroughness given to the cultural impact*
One of the biggest drawbacks against the implementation of the CIE systems in multidisciplinary sectors such as the construction industry is the cultural impact, which is taken into consideration right from the beginning in the proposed RE process, and cultural models are developed to address the cultural issues. In the development of the work redesign through process modelling, cultural models are taken into account in the visioning and the process modelling to compromise the cultural impact upon the business process re-engineering. Another cultural impact is on the implementation of the RE process in the CIE development. To reduce the cultural impact, the project manager or requirements	

(*continued*)

Table 9.1 (*Continued*)

Key issues	The coverage of the key issues
manager should stress the proper adoption of the RE process in the system development. In general, the current attitude in software development will pass through to the user requirements phase, meaning that the risk factors in coding and development will manifest itself in development failures as mentioned in Chapter 2. However, to mitigate these risk factors, the project team should properly adopt the RE process in the CIE development, which can be done through successful project and requirements management. If requirements management is conducted with product and process awareness, as instructed in the proposed RE framework, the proposed RE process can also facilitate the mitigation of the cultural impact on the RE process.	
K30,K31,K38	*Involvement of the stakeholders in the RE process*
Involvement of the stakeholders is broadly considered in the proposed RE process and they are included in nearly every phase, including elicitation, analysis, modelling and validation. Their involvement at the right time in the right subjects is important to drive user-oriented development. This is because reviews and feedback from practitioners show that direct and regular interaction with the practitioners is one of the key factors for successful acceptance and exploitation of the CIE systems in multidisciplinary sectors such as construction.	
K33,K36,K42	*Bridging the requirements with the system design*
Process modelling incorporates the whole user requirements and indicates the redesigned business process. Process modelling is sequential, while the system design is structural. It is not enough to start identifying the architecture of the system from the process model. It has the potential for misunderstanding or to miss the requirements or lack of fit between the requirements and system architecture. Therefore the requirements and system design should be bridged. To bridge the requirements with the system design, the transition from sequential thinking to structural thinking is conducted in three steps. The first transition is from the process model to the UED, which is structural. Following that is the transition from UED to use case modelling, which is sequential, and lastly, the third transition is from use case modelling to object modelling, which is structural. In each transition, requirements are transformed into system design step by step with further detail.	
K39,K44,K45	*The coverage of cost–benefits analysis and cost effectiveness*
Cost–benefits analysis is for return on investment purposes. This will analyse the estimated cost of the RE process and the whole development, such as the amount of effort required, the completion cost of the CIE system project constraints and risk analysis. The boundaries of the system to be developed and the boundaries of the project are all taken into consideration from the very beginning in the	

Table 9.1 (*Continued*)

Key issues	The coverage of the key issues
proposed RE process. Furthermore, these issues are included in the first deliverable of the requirements specification, and they are updated if required in the subsequent deliverables for change management and versioning.	
K43,K46	*Maturity of the stakeholders involved the RE processes*
It is very likely that the practitioners and stakeholders who are going to be involved in the CIE system development will be unfamiliar with the RE techniques and the system development; therefore they will not be able to follow the process and provide continuous contribution to its development. Providing the proposed RE framework in advance will enable the less knowledgeable stakeholders to be aware of what is going to happen throughout the RE process. Accordingly, stakeholders will be aware of making a more efficient and productive contribution to the development in each phase of the proposed RE process. Providing the requirements deliverables to the stakeholders will also help the stakeholders follow the progress in the RE process. At the project start-off stage, the project steering committee is reminded to define the best stakeholder groups that are knowledgeable enough to efficiently contribute to the RE process. Lastly, maturity of the stakeholders can be increased by instructions and training through requirements management.	

Table 9.2 External (independent) evaluation.

(1) Define a standard document structure

The document structures are defined but they changes in each deliverable. This is because the documents reflect the progress of the RE process, and the purpose and nature of the activities is differentiated in each stage of the proposed RE process. However, at the end of the UED, requirements specification comes to fruition, and a standard document structure is defined, which actually forms the structure of the rest of the documents in the proposed RE process.

(2) Make the document easy to change

It is adopted to ensure that the coherent association between the subsequent deliverables is established to reflect the progress and the results at the corresponding stages of the RE process, and to review the previous deliverables and write down the changes and updates of the relevant requirements in the requirements specification, referencing to the previous corresponding deliverables. Therefore, documenting strategy supports traceability, versioning and updating the document. Documentation of requirements is seen as a way of communication among the end users, managers, developers and so on.

(*continued*)

Table 9.2 (*Continued*)

(3) Uniquely identify each requirement
Involvement of the end users at every phase of the proposed RE process and approaching the user issues from discrete perspectives and continuous analysis and processing of the requirements will help to deal with each requirement.
(4) Define policies for requirements management
The main policy for requirements management is also defined. Requirements management begins with baseline requirements documentation after the project start-off phase. It involves changing and updating the requirements under control during the rest of the process in the proposed RE process. While requirements development progresses through the RE process, requirements management is continued with product and process awareness. In other words, requirements management is approached from the viewpoint of processes and techniques. Subsequently, basic techniques for requirements management include unique identification, versioning, diagnosing and change management.
(5) Define standard templates for requirements descriptions
Standard templates for requirements descriptions is adopted. While requirements activities progress through the proposed RE process, the description and features and attributes of the requirements will improve and change. The requirements deliverables describe the requirements based on the progress made on the proposed RE process. Therefore, for each requirements deliverable, a standard structure is defined in the proposed RE process.
(6) Use language simply, consistently and concisely
Simple, consistent and concise language is adopted and is one of the priority issues in the development of the proposed RE process because it is required for the stakeholders from the construction industry, who are likely not to very familiar with the RE issues. Therefore, the language used in the RE activities and that used in the production of the requirements specification deliverables are encouraged to be simple, consistent and concise because the stakeholders are involved in most of the stages of the proposed RE process.
(7) Organise formal requirements inspections
In the proposed RE process, formal inspection is only adopted at the implementation level of the system design through reverse engineering, which actually concerns the developers rather than the stakeholders and requirements engineers. In reverse engineering, the programmers generate object models from the code they have written so that they can check the original object models with those generated from the code. The generated models should match the

Table 9.2 (*Continued*)

original object models in terms of the user requirements. This is because the original object models are developed through the RE process. On the other hand, informal inspections such as informal reviews and walk-throughs are adopted. Unlike the formal inspections, informal inspections can be understood by the end users easily and then it will be easier to communicate and establish an agreed and shared understanding between the developers and the end users.

(8) Define validation checklists

The requirements validation is done through prototyping with the end user black-box tests. In these tests, requirements described in the test plans are checked with the system functionalities and the quality experienced by the requirements engineer and the end users from a multidisciplinary background. The validation is conducted iteratively until the test objectives are fulfilled in each testing phase. In addition, before the implementation stage in the development, all the requirements specification deliverables produced after the consecutive stages of the proposed RE process are provided to the end users for their views and approval. Based on the shared understanding established, the end users check the requirements and reach a consensus on the requirements. This also serves as verification and validation of the requirements before putting the system through testing.

(9) Use checklists for requirements analysis

As mentioned under the earlier criteria, each deliverable produced is distributed to all the stakeholders to check and approve the results of the corresponding phases. Furthermore, a recommended structure for the requirements deliverable is also defined, which is simple, concise and coherent and associated with each other for traceability and versioning. Furthermore, checklists are also recommended at the last section of building the shared understanding, which is a communication to the stakeholders, in the requirements analysis.

(10) Plan for conflicts and conflict resolution

Conflicts are highly expected in the development of CIE systems because it almost incorporates all the different parties from multidisciplinary sectors such as the construction industry. Therefore, the stakeholders involved in CIE developments come from different backgrounds and professions and they will inevitably have different perspectives and understanding for the user requirements. As a result, there will be conflicts in the RE process. These conflicts are addressed by building a shared understanding, communication and active collaboration with the end users and negotiation, brainstorming, workshops and review meetings in the proposed RE process.

Table 9.3 External evaluation of the project risk factors.

Risk factor	The proposed RE process to mitigate the risk factors
Misunderstanding the requirements	1. Documenting requirements 2. Interacting with the stakeholders 3. Reviewing requirements documentation and requirements 4. Process modelling and product modelling 5. Prototyping with the end users tests
Lack of adequate user involvement	Using techniques that require the involvement of the users 1. Interviews 2. Meetings 3. Work modelling 4. Workshops 5. Process modelling 6. UED 7. Use case modelling 8. User tests, etc.
Failure to manage end user expectations	1. Interacting with users 2. Documenting requirements 3. Prioritising requirements 4. Validating requirements with early prototypes
Changing scope/ objections	1. Documenting business goals 2. Context and requirements 3. Requirements management 4. Change requests
Lack of frozen requirements	1. Documenting business goals and objectives 2. Requirements management 3. Documenting requirements 4. Change management
Conflict between the user departments	1. Building a shared understanding 2. Consolidation of work models 3. Affinity diagrams 4. Process modelling with storyboarding 5. UED 6. Use case modelling 7. Documenting requirements 8. Meetings with the stakeholders 9. Reviewing requirements documents and requirements 10. Prototyping with end user tests

Table 9.3 (*Continued*)

Risk factor	The proposed RE process to mitigate the risk factors
Incomplete requirements and specification	1. Documenting requirements 2. Work modelling 3. Text-based storyboarding for process modelling 4. UED 5. Consolidation of work models and affinity diagrams 6. Communication to the stakeholders 7. Reviewing the requirements documents and requirements 8. End user tests for validation
Ambiguous and vague requirements	1. Documenting requirements 2. Building shared understanding 3. Process modelling 4. UED 5. Use case modelling 6. Requirements validation

It shows that there are precautions in the proposed RE process to eliminate the risk factors identified in Chapter 2. Consequently, the external evaluation of the proposed RE framework through the use of independent sources such as the REAIMS assessment criteria and RE-related project risk factors is then completed.

References

Nikula, U., (2003), "Experiences from Lightweight RE Method Evaluations", IEEE International Workshop of Comparative Evaluation in Requirements Engineering, Monterey Bay, California, IEEE Computer Society, pp. 53 60.

10

Summary and Conclusion

10.1 Introduction

The scope and the approach adopted from various computer integrated environment (CIE) development experiences in the last decade focus on requirements engineering (RE) to provide a means for user-oriented and practical CIE systems developments for multidisciplinary sectors. In order to provide a comprehensive ready-to-use RE framework for the successful user-driven developments and to contribute towards the effective implementation of the CIE systems the RE practices from Case Studies 3 and 4, which had high awareness about aligning technology with people and processes, were evaluated and analysed using the qualitative and quantitative scientific techniques. This concluding chapter summarises the book and evaluates the implications of the proposed RE framework.

10.2 Contribution to Knowledge Society

The contribution of this book is the framework of the RE process depicted in Figure 10.1 in the development of CIE for multidisciplinary sectors such as the construction industry. This framework has been developed and proposed after many years of research and experience in the area of CIE developments. In this book, the framework development and explanation have included a prescription of the RE process, its limitations, key features, constructs, usage and documentation. The proposed RE approach was justified by the availability of a complete method description. The key characteristics of the method can be summarised as follows:

- Ready to use
- Simple
- Domain specific
- Adaptable

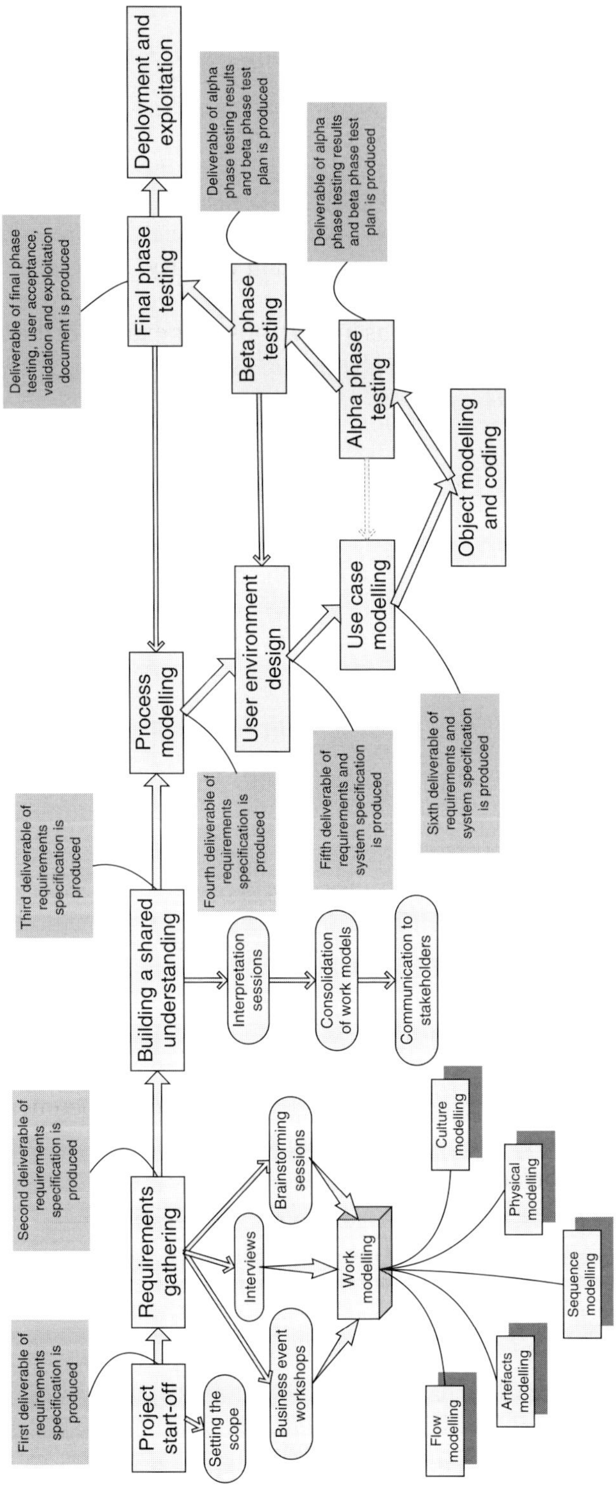

Figure 10.1 The proposed requirements engineering framework.

- Systematic
- Integration with the legacy systems.

From these key characteristics, the following three key constructs were derived for the proposed framework:

- Requirements development and improvement activities
- Requirements document templates
- Facilitating requirements management activity.

However, the framework has limitations. For example, the above key constructs were not clearly supplemented by two additional constructs. These are as follows:

- Tool support for requirements management and
- Training.

The proposed RE framework provided a balanced approach to the different areas of the RE process. Evaluation against the key issues did not present any big surprises since the proposed framework was designed with these key issues in mind. Furthermore, the evaluation against the top 10 guidelines and project risk factors of the Requirements Engineering Adaptation and Improvement for Safety and dependability (REAIMS) found that neither presented any big surprises.

As a result, based on the evaluation and assessment of the proposed framework in Chapter 9, the proposed RE framework appears to be ready for use and provides the basic artefacts required for RE to start developing requirements and managing requirements for a CIE development project targeted at the multidisciplinary sectors. However, the weakest part of the framework is clearly the tool support for requirements management and training. Requirements management is currently limited to the requirements specifications documents. However, there are freeware and commercial tools available to complement this obvious shortage.

Secondly, in early 2000, researchers in construction information technology acknowledged that the issues of integration, communication, networking and collaboration will be popular topics of the construction industry (Alshawi *et al.*, 1998; Sarshar *et al.*, 2000; Sarshar *et al.*, 2002); nowadays, it is clearly evident that building information modelling (BIM) as a computer integrated environment for the construction industry, which emerged almost 30 years ago, has become a very popular concept, and there is a demand for effective implementation in the construction industry. Furthermore, BIM has now become a mandated requirement for public property projects in some countries such as Denmark. This is because change management is the main topic that will look after the transition for BIM adoption and implementation in the construction sector. The proposed RE framework is an enabler for successful change management. In other words, effective and correct RE will be vital for the development of the technology and the effective uptake of BIM in the construction industry. Therefore, this book will lay the foundation for the best practices of the RE research and implementation of BIM in the construction industry.

10.3 Main Conclusions

The following are the initial conclusions drawn from this research effort.

1. Although there is a slow increasing awareness about RE in the CIE developments for multidisciplinary sectors, such as construction, the CIE concept BIM has gained a lot of importance. Furthermore, so far there has not been any other RE framework for BIM implementation for computer integrated construction via the CIE systems.
2. In order for the development of the CIE technologies to contribute to the effective uptake of these technologies by the multidisciplinary sectors, there is a need to define a framework of RE.
3. This book has elaborated and provided a framework of RE for the CIE developments for the multidisciplinary sectors, using some case study projects. The RE process is depicted in Figure 10.1.
4. A ready-to-use method of RE process must address all the key areas of the problem domain. These key areas should be processed through the RE activities such as requirements elicitation, requirements analysis and modelling and requirements validation.
5. A ready-to-use method must be construction IT domain specific to allow for the three constructs (*requirements development and improvement activities, requirements document templates, facilitating requirements management activity)* that fit the domain without modification.
6. A shared and common understanding of the points given below is important for the successful implementation of RE:
   - CIE development
   - User requirements and needs
   - Interpretation of tangible and intangible artefacts and objects used in the multidisciplinary sector

- Communication and collaboration between the project team
- Knowledge and its transformation
- Requirements documentation for requirements management.

7. In the development and proposition of the RE framework, the barriers against the implementation of the CIE systems such as process issues, cultural issues, people attitudes, fragmentation, etc. are also taken into consideration in order to mitigate their impacts in the development of the technology and in the uptake of the system by multidisciplinary sectors.

Although the proposed RE framework has been improved and proposed from experience and research, it should be adopted in future development and implementations in order to see its effectiveness and usefulness in practice. Afterwards, the maturity and usefulness of the RE process can be improved further towards level 2 and level 3, by measuring the maturity of the method by means of the capability maturity models, such as the REAIMS model.

The proposed framework includes activities for requirements elicitation, requirements analysis, requirements modelling and requirements validation. Furthermore, the framework also facilitates the requirements management by providing useful techniques. User involvement, building a shared understanding, communication and collaboration and prototyping can be regarded as key corner stones of the framework.

10.4 Recommendations for the Future

The main recommendations and future research areas can be the improvement of the quality of RE services, the quality of the CIE systems and cost-related issues in order to enhance the development of the CIE technology and its effective implementation in the construction industry.

1. The framework can be implemented in practice in the future CIE developments in both commercial and research-based projects so that the framework can be further improved or alternative solutions can be considered.
2. In addition to three key constructs that were covered in the proposed RE framework, the two supplementary key constructs, which are training and tool support for requirements management, can also be investigated to further the communication, collaboration and shared understanding amongst the CIE development team.

Figure 10.1 is the RE process and techniques to be used for the CIE systems development for multidisciplinary sectors such as construction, particularly for BIM implementation and adoption in the construction industry.

References

Alshawi, M., Aouad, G., Faraj, I., Child, T., Underwood, J., (1998), "The implementation of the Industry Foundation Classes in integrated environments", CIB W78 Conference, Sweden, August, 1998, pp. 55–66.

Sarshar, M., Betts, M., Abbott, C., Aouad, G., (2000), "A Vision for Construction IT 2005–2010", Research Series, London, United Kingdom, RICS (Royal Institute of Chartered Surveyors), pp. 1–42.

Sarshar, M., Tanyer, M., Aouad, G., Underwood, J., (2002), "A Vision for Construction IT 2005–2010: Two Case Studies", Engineering, Construction and Architectural Management, 9(2):152–160.

Index